CHAPTER 1

First-Order Differential Equations

EXERCISES FOR SECTION 1.1

1. **(a)** The equilibrium solutions correspond to the values of P for which $dP/dt = 0$ for all t. For this equation, $dP/dt = 0$ for all t if $P = 0$ or $P = 230$.
   (b) The population is increasing if $dP/dt > 0$. That is, $P(1 - P/230) > 0$. Hence, $0 < P < 230$.
   (c) The population is decreasing if $dP/dt < 0$. That is, $P(1 - P/230) < 0$. Hence, $P > 230$ or $P < 0$. Since this is a population model, $P < 0$ might be considered "nonphysical."

3. In order to answer the question, we first need to analyze the sign of the polynomial $y^3 - y^2 - 12y$. Factoring, we obtain

$$y^3 - y^2 - 12y = y(y^2 - y - 12) = y(y - 4)(y + 3).$$

   (a) The equilibrium solutions correspond to the values of y for which $dy/dt = 0$ for all t. For this equation, $dy/dt = 0$ for all t if $y = -3$, $y = 0$, or $y = 4$.
   (b) The solution $y(t)$ is increasing if $dy/dt > 0$. That is, $-3 < y < 0$ or $y > 4$.
   (c) The solution $y(t)$ is decreasing if $dy/dt < 0$. That is, $y < -3$ or $0 < y < 4$.

5. The rate of learning is dL/dt. Thus, we want to know the values of L between 0 and 1 for which dL/dt is a maximum. As $k > 0$ and $dL/dt = k(1 - L)$, dL/dt attains it maximum value at $L = 0$.

7. **(a)** We have $L_B(0) = L_J(0) = 0$. So Jillian's rate of learning at $t = 0$ is dL_J/dt evaluated at $t = 0$. At $t = 0$, we have

$$\frac{dL_J}{dt} = 2(1 - L_J) = 2.$$

   Beth's rate of learning at $t = 0$ is

$$\frac{dL_B}{dt} = 3(1 - L_B)^2 = 3.$$

   Hence Beth's rate is larger.
   (b) In this case, $L_B(0) = L_J(0) = 1/2$. So Jillian's rate of learning at $t = 0$ is

$$\frac{dL_J}{dt} = 2(1 - L_J) = 1$$

   because $L_J = 1/2$ at $t = 0$. Beth's rate of learning at $t = 0$ is

$$\frac{dL_B}{dt} = 3(1 - L_B)^2 = \frac{3}{4}$$

   because $L_B = 1/2$ at $t = 0$. Hence Jillian's rate is larger.
   (c) In this case, $L_B(0) = L_J(0) = 1/3$. So Jillian's rate of learning at $t = 0$ is

$$\frac{dL_J}{dt} = 2(1 - L_J) = \frac{4}{3}.$$

   Beth's rate of learning at $t = 0$ is

$$\frac{dL_B}{dt} = 3(1 - L_B)^2 = \frac{4}{3}.$$

   They are both learning at the same rate when $t = 0$.

Student Solutions Manual for

DIFFERENTIAL EQUATIONS

second edition

Paul Blanchard,
Robert L. Devaney,
and Glen R. Hall

Boston University

THOMSON
™
BROOKS/COLE

Australia • Canada • Mexico • Singapore • Spain • United Kingdom • United States

Printed in Canada
1 2 3 4 5 6 7   05 04 03 02 01

Printer: Webcom Limited

ISBN: 0-534-38516-8

For more information about our products,
contact us at:
Thomson Learning Academic Resource Center
1-800-423-0563

For permission to use material from this text,
contact us by:
Phone: 1-800-730-2214
Fax: 1-800-731-2215
Web: http://www.thomsonrights.com

Asia
Thomson Learning
5 Shenton Way #01-01
UIC Building
Singapore 068808

Australia
Nelson Thomson Learning
102 Dodds Street
South Street
South Melbourne, Victoria 3205
Australia

Canada
Nelson Thomson Learning
1120 Birchmount Road
Toronto, Ontario M1K 5G4
Canada

Europe/Middle East/South Africa
Thomson Learning
High Holborn House
50/51 Bedford Row
London WC1R 4LR
United Kingdom

Latin America
Thomson Learning
Seneca, 53
Colonia Polanco
11560 Mexico D.F.
Mexico

Spain
Paraninfo Thomson Learning
Calle/Magallanes, 25
28015 Madrid, Spain

CONTENTS

9. The general solution of the differential equation $dr/dt = -\lambda r$ is $r(t) = r_0 e^{-\lambda t}$ where $r(0) = r_0$ is the initial amount.

(a) We have $r(t) = r_0 e^{-\lambda t}$ and $r(5230) = r_0/2$. Thus

$$\frac{r_0}{2} = r_0 e^{-\lambda \cdot 5230}$$

$$\frac{1}{2} = e^{-\lambda \cdot 5230}$$

$$\ln \frac{1}{2} = -\lambda \cdot 5230$$

$$-\ln 2 = -\lambda \cdot 5230$$

because $\ln 1/2 = -\ln 2$. Thus,

$$\lambda = \frac{\ln 2}{5230} \approx 0.000132533.$$

(b) We have $r(t) = r_0 e^{-\lambda t}$ and $r(8) = r_0/2$. By a computation similar to the one in part (a), we have

$$\lambda = \frac{\ln 2}{8} \approx 0.0866434.$$

(c) If $r(t)$ is the number of atoms of C-14, then the units for dr/dt is number of atoms per year. Since $dr/dt = -\lambda r$, λ is "per year." Similarly, for I-131, λ is "per day." The unit of measurement of r does not matter.

(d) We get the same answer because the original quantity, r_0, cancels from each side of the equation. We are only concerned with the proportion remaining (one-half of the original amount).

11. The most important difference between Carbon-14 and Carbon-12 is that C-14 is radioactive. The models we constructed above are based on the assumption that the number of C-14 atoms that decay in a particular time period is proportional to the number of C-14 atoms present where the constant of proportionality λ is determined from the half-life. Hence we can determine the amount of C-14 in a sample by counting the number of atoms that decay in a given time period. Divide this number by the time elapsed to obtain an approximation for dr/dt. Then, using $dr/dt = -\lambda r$, divide by λ to get r (making sure the units agree).

13. Let $P(t)$ be the population at time t, k be the growth-rate parameter, and N be the carrying capacity. The modified models are

(a) $dP/dt = k(1 - P/N)P - 100$
(b) $dP/dt = k(1 - P/N)P - P/3$
(c) $dP/dt = k(1 - P/N)P - a\sqrt{P}$, where a is a positive parameter.

15. Several different models are possible. Let $R(t)$ denote the rhinoceros population at time t. The basic assumption is that there is a minimum threshold that the population must exceed if it is to survive. In terms of the differential equation, this assumption means that dR/dt must be negative if R is close to zero. Three models that satisfy this assumption are:

- If k is a growth-rate parameter and M is a parameter measuring when the population is "too small", then

$$\frac{dR}{dt} = kR\left(\frac{R}{M} - 1\right).$$

- If k is a growth-rate parameter and b is a parameter that determines the level the population will start to decrease ($R < b/k$), then

$$\frac{dR}{dt} = kR - b.$$

- If k is a growth-rate parameter and b is a parameter that determines the extinction threshold, then

$$\frac{dR}{dt} = aR - \frac{b}{R}.$$

In each case, if R is below a certain threshold, dR/dt is negative. Thus, the rhinos will eventually die out. The choice of which model to use depends on other assumptions. There are other equations that are also consistent with the basic assumption.

17. **(a)** Since the population seems to level off at 64, we try a logistic model. Let t be the time measured in years and $P(t)$ be the population. The model is then

$$\frac{dP}{dt} = kP\left(1 - \frac{P}{N}\right)$$

where k is the growth-rate parameter and N is the carrying capacity.

(b) Take $N = 64$ as the carrying capacity. This is reasonable since the population has "leveled off" for the four years 1956–1959. To compute k we could use the same technique as we did for the U.S. population, but since we already know the carrying capacity and since $P(0) = 34$ is a significant fraction of the carrying capacity, it is better to use the differential equation to determine k. To approximate dP/dt at time $t = 0$ (which we take to be 1947), we calculate the slope of the line connecting the first two data points $(0, 34)$ and $(1, 40)$. The slope of this line gives us an approximation of dP/dt at $t = 0$. That is,

$$\frac{40 - 34}{1 - 0} \approx \left.\frac{dP}{dt}\right|_{t=0} = kP(0)\left(1 - \frac{P(0)}{64}\right),$$

so

$$6 = k \cdot 34\left(1 - \frac{34}{64}\right),$$

which yields $k \approx 0.38$.

(c) Since the population is at the carrying capacity in 1959, it will remain there, so we predict that the population today is 64.

19. **(a)** We consider dx/dt in each system. Setting $y = 0$ yields $dx/dt = 5x$ in system (i) and $dx/dt = x$ in system (ii). If the number x of prey is equal for both systems, dx/dt is larger in system (i). Therefore, the prey in system (i) reproduce faster if there are no predators.

(b) We must see what affect the predators (represented by the y-terms) have on dx/dt in each system. Since the magnitude of the coefficient of the xy-term is larger in system (ii) than in system (i), y has a greater effect on dx/dt in system (ii). Hence the predators have a greater effect on the rate of change of the prey in system (ii).

(c) We must see what affect the prey (represented by the x-terms) have on dy/dt in each system. Since x and y are both nonnegative, it follows that

$$-2y + \tfrac{1}{2}xy < -2y + 6xy,$$

and therefore, if the number of predators is equal for both systems, dy/dt is smaller in system (i). Hence more prey are required in system (i) than in system (ii) to achieve a certain growth rate.

21. **(a)** The independent variable is t, and x and y are dependent variables. Since each xy-term is positive, the presence of either species increases the rate of change of the other. Hence, these species cooperate. The parameter α is the growth-rate parameter for x, and γ is the growth-rate parameter for y. The parameter N represents the carrying capacity for x, but y has no carrying capacity. The parameter β measures the benefit to x of the interaction of the two species, and δ measures the benefit to y of the interaction.

(b) The independent variable is t, and x and y are the dependent variables. Since both xy-terms are negative, these species compete. The parameter γ is the growth-rate coefficient for x, and α is the growth-rate parameter for y. Neither population has a carrying capacity. The parameter δ measures the harm to x caused by the interaction of the two species, and β measures the harm to y caused by the interaction.

EXERCISES FOR SECTION 1.2

1. **(a)** Let's check Bob's solution first. Since $dy/dt = 1$ and

$$\frac{y(t) + 1}{t + 1} = \frac{t + 1}{t + 1} = 1,$$

Bob's answer is correct.

Now let's check Glen's solution. Since $dy/dt = 2$ and

$$\frac{y(t) + 1}{t + 1} = \frac{2t + 2}{t + 1} = 2,$$

Glen's solution is also correct.

Finally let's check Paul's solution. We have $dy/dt = 2t$ on one hand and

$$\frac{y(t) + 1}{t + 1} = \frac{t^2 - 1}{t + 1} = t - 1$$

on the other. Therefore, Paul is wrong.

(b) At first glance, they should have seen the equilibrium solution $y(t) = -1$ for all t because $dy/dt = 0$ for any constant function and $y = -1$ implies that

$$\frac{y + 1}{t + 1} = 0$$

independent of t.

Strictly speaking the differential equation is not defined for $t = -1$, and hence the solutions are not defined for $t = -1$.

3. In order to find one such $f(t, y)$, we compute the derivative of $y(t)$. We obtain

$$\frac{dy}{dt} = \frac{de^{t^3}}{dt} = 3t^2 e^{t^3}.$$

Now we replace e^{t^3} in the last expression by y and get the differential equation

$$\frac{dy}{dt} = 3t^2 y.$$

5. The constant function $y(t) = 0$ is an equilibrium solution.
For $y \neq 0$ we separate the variables and integrate

$$\int \frac{dy}{y} = \int t \, dt$$

$$\ln |y| = \frac{t^2}{2} + c$$

$$|y| = c_1 e^{t^2/2}$$

where $c_1 = e^c$ is an arbitrary positive constant.
If $y > 0$, then $|y| = y$ and we can just drop the absolute value signs in this calculation. If $y < 0$, then $|y| = -y$, so $-y = c_1 e^{t^2/2}$. Hence, $y = -c_1 e^{t^2/2}$. Therefore,

$$y = k e^{t^2/2}$$

where $k = \pm c_1$. Moreover, if $k = 0$, we get the equilibrium solution. Thus, $y = k e^{t^2/2}$ yields all solutions to the differential equation if we let k be any real number. (Strickly speaking we need a theorem from Section 1.5 to justify the assertion that this formula provides all solutions.)

7. We separate variables and integrate to obtain

$$\int \frac{dy}{2y + 1} = \int dt.$$

We get

$$\frac{1}{2} \ln |2y + 1| = t + c$$

$$|2y + 1| = c_1 e^{2t},$$

where $c_1 = e^{2c}$. As in Exercise 5, we can drop the absolute value signs by replacing $\pm c_1$ with a new constant k_1. Hence, we have

$$2y + 1 = k_1 e^{2t}$$

$$y = \frac{1}{2} \left(k_1 e^{2t} - 1 \right),$$

and letting $k = k_1/2$, $y(t) = k e^{2t} - 1/2$. Note that, for $k = 0$, we get the equilibrium solution.

9. We separate variables and integrate to obtain

$$\int e^y \, dy = \int dt$$

$$e^y = t + c,$$

where c is any constant. We obtain $y(t) = \ln(t + c)$.

11. First note that the differential equation is not defined if $y = 0$.
In order to separate the variables, we write the equation as

$$\frac{dy}{dt} = \frac{t}{y(t^2 + 1)}$$

to obtain

$$\int y \, dy = \int \frac{t}{t^2 + 1} \, dt$$

$$\frac{y^2}{2} = \frac{1}{2} \ln(t^2 + 1) + c,$$

where c is any constant. So we get

$$y^2 = \ln\left(k(t^2 + 1)\right),$$

where $k = e^{2c}$ (hence any positive constant). We have

$$y(t) = \pm\sqrt{\ln\left(k(t^2 + 1)\right)},$$

where k is any positive constant and the sign is determined by the initial condition.

13. First note that the differential equation is not defined for $y = -1/2$. We separate variables and integrate to obtain

$$\int (2y + 1) \, dy = \int dt$$

$$y^2 + y = t + k,$$

where k is any constant. So

$$y(t) = \frac{-1 \pm \sqrt{4t + 4k + 1}}{2} = \frac{-1 \pm \sqrt{4t + c}}{2},$$

where c is any constant and the $\pm$ sign is determined by the initial condition.
We can rewrite the answer in the more simple form

$$y(t) = -\frac{1}{2} \pm \sqrt{t + c_1}$$

where $c_1 = k + 1/4$. If k can be any possible constant, then c_1 can be as well.

15. First of all, the equilibrium solutions are $y = 0$ and $y = 1$. Now suppose $y \neq 0$ and $y \neq 1$. We separate variables to obtain

$$\int \frac{1}{y(1-y)} \, dy = \int dt = t + c,$$

where c is any constant. To integrate, we use partial fractions. Write

$$\frac{1}{y(1-y)} = \frac{A}{y} + \frac{B}{1-y}.$$

We must have $A = 1$ and $-A + B = 0$. Hence, $A = B = 1$ and

$$\frac{1}{y(1-y)} = \frac{1}{y} + \frac{1}{1-y}.$$

Consequently,

$$\int \frac{1}{y(1-y)} \, dy = \ln|y| - \ln|1 - y| = \ln \left| \frac{y}{1-y} \right|.$$

After integration, we have

$$\ln \left| \frac{y}{1-y} \right| = t + c$$

$$\left| \frac{y}{1-y} \right| = c_1 e^t,$$

where $c_1 = e^c$ is any positive constant. To remove the absolute value signs, we replace the positive constant c_1 with a constant k that can be any real number and get

$$y(t) = \frac{ke^t}{1 + ke^t},$$

where $k = \pm c_1$. If $k = 0$, we get the first equilibrium solution. The formula $y(t) = ke^t/(1 + ke^t)$ yields all the solutions to the differential equation except for the equilibrium solution $y(t) = 1$.

17. The equation can be written in the form

$$\frac{dy}{dt} = (y + 1)(t^2 + 1).$$

From this, we see that there is an equilibrium solution where $y = -1$. Separating variables and integrating, we obtain

$$\int \frac{dy}{y+1} = \int t^2 + 1 \, dt$$

$$\ln|y + 1| = \frac{t^3}{3} + t + c,$$

where c is any constant. Thus,

$$|y + 1| = c_1 e^{t + t^3/3},$$

where $c_1 = e^c$. We can dispose of the absolute value signs by allowing the constant c_1 to be any real number. In other words,

$$y(t) = -1 + ke^{t+t^3/3},$$

where $k = \pm c_1$. Note that, if $k = 0$, we have the equilibrium solution.

19. The function $y(t) = 0$ for all t is an equilibrium solution.
   Suppose $y \neq 0$ and separate variables. We get

$$\int y + \frac{1}{y} \, dy = \int e^t \, dt$$

$$\frac{y^2}{2} + \ln |y| = e^t + c,$$

where c is any real constant. We cannot solve this equation for y, so we leave the expression for y in this implicit form. Note that the equilibrium solution $y = 0$ cannot be obtained from this implicit equation.

21. The constant function $w(t) = 0$ is an equilibrium solution. Suppose $w \neq 0$ and separate variables. We get

$$\int \frac{dw}{w} = \int \frac{dt}{t}$$

$$\ln |w| = \ln |t| + c$$

$$= \ln c_1 |t|,$$

where c is any constant and $c_1 = e^c$. Therefore,

$$|w| = c_1 |t|.$$

We can eliminate the absolute value signs by allowing the constant to assume positive or negative values. We have

$$w = kt,$$

where $k = \pm c_1$. Moreover, if $k = 0$ we get the equilibrium solution.

23. We separate variables to obtain

$$\int (y^4 + 3y) \, dy = \int (t^2 + 1) \, dt.$$

Integrating, we have

$$\frac{y^5}{5} + \frac{3y^2}{2} = \frac{t^3}{3} + t + c,$$

where c is any real constant. We cannot solve the above equation for y, so we leave the solution in this implicit form.

25. From Exercise 7, we already know that the general solution is

$$y(t) = ke^{2t} - \tfrac{1}{2},$$

so we need only find the constant k for which $y(0) = 3$. We solve

$$3 = ke^0 - \tfrac{1}{2}$$

for k and obtain $k = 7/2$. The solution of the initial-value problem is

$$y(t) = \tfrac{7}{2}e^{2t} - \tfrac{1}{2}.$$

27. Separating variables and integrating, we obtain

$$\int \frac{dy}{y^2} = -\int dt$$

$$-\frac{1}{y} = -t + c.$$

So we get

$$y = \frac{1}{t - c}.$$

Now we need to find the constant c so that $y(0) = 1/2$. To do this we solve

$$\frac{1}{2} = \frac{1}{0 - c}$$

and get $c = -2$. The solution of the initial-value problem is

$$y(t) = \frac{1}{t + 2}.$$

29. We do not need to do any computations to solve this initial-value problem. We know that the constant function $y(t) = 0$ for all t is an equilibrium solution, and it satisfies the initial condition.

31. We can write the equation in the form

$$\frac{dy}{dt} = \frac{t^2}{y(t^3 + 1)}$$

and separate variables to obtain

$$\int y \, dy = \int \frac{t^2}{t^3 + 1} \, dt$$

$$\frac{y^2}{2} = \frac{1}{3} \ln |t^3 + 1| + c,$$

where c is a constant. Hence,

$$y^2 = \frac{2}{3} \ln |t^3 + 1| + 2c.$$

The initial condition $y(0) = -2$ implies

$$4 = (-2)^2 = \frac{2}{3} \ln |1| + 2c.$$

Thus, $c = 2$, and

$$y(t) = -\sqrt{\frac{2}{3} \ln(t^3 + 1) + 4}.$$

We choose the negative square root because $y(0)$ is negative.

33. We separate variables to obtain

$$\int \frac{dy}{1 + y^2} = \int t \, dt$$

$$\arctan y = \frac{t^2}{2} + c,$$

where c is a constant. Hence the general solution is

$$y(t) = \tan\left(\frac{t^2}{2} + c\right).$$

Next we find c so that $y(0) = 1$. Solving

$$1 = \tan\left(\frac{0^2}{2} + c\right)$$

yields $c = \pi/4$, and the solution to the initial-value problem is

$$y(t) = \tan\left(\frac{t^2}{2} + \frac{\pi}{4}\right).$$

35. Let $S(t)$ denote the amount of salt (in pounds) in the bucket at time t (in minutes). We derive a differential equation for S by considering the difference between the rate that salt is entering the bucket and the rate that salt is leaving the bucket. Salt is entering the bucket at the rate of 1/4 pounds per minute. The rate that salt is leaving the bucket is the product of the concentration of salt in the mixture and the rate that the mixture is leaving the bucket. The concentration is $S/5$, and the mixture is leaving the bucket at the rate of 1/2 gallons per minute. We obtain the differential equation

$$\frac{dS}{dt} = \frac{1}{4} - \frac{S}{5} \cdot \frac{1}{2},$$

which can be rewritten as

$$\frac{dS}{dt} = \frac{5 - 2S}{20}.$$

This differential equation is separable, and we can find the general solution by integrating

$$\int \frac{1}{5 - 2S}\, dS = \int \frac{1}{20}\, dt.$$

We have

$$-\frac{\ln|5 - 2S|}{2} = \frac{t}{20} + c$$

$$\ln|5 - 2S| = -\frac{t}{10} + c_1$$

$$|5 - 2S| = c_2 e^{-t/10}.$$

We can eliminate the absolute value signs and determine c_2 using the initial condition $S(0) = 0$ (the water is initially free of salt). We have $c_2 = 5$, and the solution is

$$S(t) = 2.5 - 2.5e^{-t/10} = 2.5(1 - e^{-t/10}).$$

(a) When $t = 1$, we have $S(1) = 2.5(1 - e^{-0.1}) \approx 0.238$ lbs.

(b) When $t = 10$, we have $S(10) = 2.5(1 - e^{-1}) \approx 1.58$ lbs.

(c) When $t = 60$, we have $S(60) = 2.5(1 - e^{-6}) \approx 2.49$ lbs.

(d) When $t = 1000$, we have $S(1000) = 2.5(1 - e^{-100}) \approx 2.50$ lbs.

(e) When t is very large, the $e^{-t/10}$ term is close to zero, so $S(t)$ is very close to 2.5 lbs. In this case, we can also reach the same conclusion by doing a qualitative analysis of the solutions of the equation. The constant solution $S(t) = 2.5$ is the only equilibrium solution for this equation, and by examining the sign of dS/dt, we see that all solutions approach $S = 2.5$ as t increases.

37. (a) If we let k denote the proportionality constant in Newton's law of cooling, the differential equation satisfied by the temperature T of the chocolate is

$$\frac{dT}{dt} = k(T - 70).$$

We also know that $T(0) = 170$ and that $dT/dt = -20$ at $t = 0$. Therefore, we obtain k by evaluating the differential equation at $t = 0$. We have

$$-20 = k(170 - 70),$$

so $k = -0.2$. The initial-value problem is

$$\frac{dT}{dt} = -0.2(T - 70), \quad T(0) = 170.$$

(b) We can solve the initial-value problem in part (a) by separating variables. We have

$$\int \frac{dT}{T - 70} = \int -0.2\, dt$$

$$\ln|T - 70| = -0.2t + k$$

$$|T - 70| = ce^{-0.2t}.$$

Since the temperature of the chocolate cannot become lower than the temperature of the room, we can ignore the absolute value and conclude

$$T(t) = 70 + ce^{-0.2t}.$$

Now we use the initial condition $T(0) = 170$ to find the constant c because

$$170 = T(0) = 70 + ce^{-0.2(0)},$$

which implies that $c = 100$. The solution is

$$T = 70 + 100e^{-0.2t}.$$

In order to find t so that the temperature is $110°$ F, we solve

$$110 = 70 + 100e^{-0.2t}$$

for t obtaining

$$\tfrac{2}{5} = e^{-0.2t}$$

$$\ln \tfrac{2}{5} = -0.2t$$

so that

$$t = \frac{\ln(2/5)}{-0.2} \approx 4.6.$$

39. Let $C(t)$ denote the concentration. From Exercise 38, the differential equation is

$$\frac{dC}{dt} = \frac{20 - 3C}{100}$$

Note that $dC/dt = 0$ for all t if $C = 20/3$. Hence, the constant function $C(t) = 20/3$ is an equilibrium solution.

Separating variables and integrating, we obtain

$$\int \frac{dC}{20 - 3C} = \int \frac{1}{100} \, dt$$

$$-\frac{1}{3} \ln |20 - 3C| = \frac{t}{100} + c$$

$$|20 - 3C| = c_1 e^{-3t/100}.$$

To remove the absolute value signs we note that $e^{-3t/100}$ is never zero, so $20 - 3C(t)$ is either positive for all t, negative for all t, or zero for all t. Hence the general solution is

$$20 - 3C = k_1 e^{-3t/100},$$

where k_1 can be any constant. Finally, we have

$$C(t) = \tfrac{20}{3} + k e^{-3t/100}$$

where k can be any constant.

41. First we find the general solution without using the specific values of the parameters. The equation

$$\frac{dM}{dt} = iM - p$$

is separable, so we separate variables and integrate to obtain

$$\int \frac{dM}{iM - p} = \int dt$$

$$\frac{1}{i} \ln |iM - p| = t + c_1$$

$$|iM - p| = c_2 e^{it}$$

where $c_2 = e^{ic_1}$. Hence,

$$iM - p = ke^{it}$$

(remember i is the interest rate, not $\sqrt{-1}$).

There is an equilibrium solution if $dM/dt = iM - p = 0$ for all t. Consequently there is a constant solution if $p = iM$ (the amount of the payments is exactly equal to the interest).

Since e^{it} is never zero, the sign of $iM - p$ never changes. Hence $iM(t) - p > 0$ for all t, $iM(t) - p < 0$ for all t, or $iM(t) - p = 0$ for all t. The general solution is

$$M = \frac{p + ke^{it}}{i},$$

where k is a constant which is determined by the initial condition. (Typically k is negative if $M(t)$ is eventually zero.)

We know that $M(0) = 150,000$ and $M(30) = 0$. Thus, we obtain the system of equations

$$\begin{cases} \dfrac{p + k}{i} = 150,000 \\ \dfrac{p + ke^{30i}}{i} = 0. \end{cases}$$

Subtracting the second equation from the first yields

$$\frac{k(1 - e^{30i})}{i} = 150,000,$$

and consequently,

$$k = \frac{150,000\, i}{1 - e^{30i}}.$$

Since

$$p + ke^{30i} = 0,$$

we have

$$p = -ke^{30i} = -\frac{150,000\, i}{1 - e^{30i}} e^{30i} = \frac{150,000\, i}{1 - e^{-30i}}.$$

Note that the total amount paid over the thirty year life of the loan is $30p$.

(a) Substituting $i = 0.07$, we get a total amount paid of $30p = \$358,957$. At $i = 0.065$, we get $30p = \$341,018$, and adding the \$4,500 extra due to points, we get a total cost of \$345,518.

(b) The 6.5% option is better.

(c) If the $4,500 is invested at 5%, then after thirty years the $4,500 has grown to $20,168 (the solution of $dy/dt = 0.05y$, $y(0) = 4,500$.) The 7% option is the better deal.

EXERCISES FOR SECTION 1.3

1.

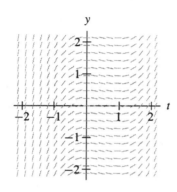

3.

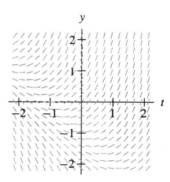

5.

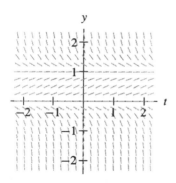

7. (a)

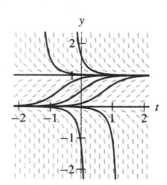

(b) The solution with $y(0) = 1/2$ approaches the equilibrium value $y = 1$ from below as t increases. It decreases toward $y = 0$ as t decreases.

9. (a)

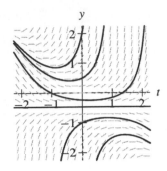

(b) The solution $y(t)$ with $y(0) = 1/2$ has $y(t) \to \infty$ both as t increases and as t decreases.

11.   (a) On the line $y = 3$ in the ty-plane, all of the slope marks have slope -1.

(b) Because f is continuous, if y is close to 3, then $f(t, y) < 0$. So any solution close to $y = 3$ must be decreasing. Therefore, solutions $y(t)$ that satisfy $y(0) < 3$ can never be larger than 3 for $t > 0$, and consequently $y(t) < 3$ for all t.

13. The slope field in the ty-plane is constant along vertical lines.

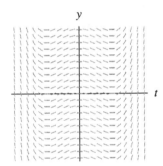

15.   (a) Note that the slopes are constant along vertical lines—lines along which t is constant, so the right-hand side of the corresponding equation must depend only on t. The two such choices are equations (i) and (iv). Because the slope is negative for $t > 1$ and positive for $t < 1$, this slope field must correspond to equation (iv).

(b) This slope field has an equilibrium solution corresponding to the line $y = 1$, so it must be correspond to either equation (ii), (v), or (viii). Both (ii) and (viii) have another equilibrium solution corresponding to $y = -1$, so this slope field must correspond to equation (v).

(c) This slope field has equilibrium solutions corresponding to $y = \pm 1$. Hence it corresponds to either equation (ii) or (viii). Since dy/dt is negative along $y = 0$, this slope field must correspond to equation (viii).

(d) This slope field depends both on y and on t, so it can only correspond to equation (iii) or (vi). When $t = 0$ the slopes are positive, so the slope field must correspond to equation (iii).

17.   (a) Because the equation is autonomous, the slope field is constant on horizontal lines, so this solution provides enough information to sketch the slope field on the entire upper half plane. Also, if we assume that f is continuous, then the slope field on the line $y = 0$ must be horizontal.

(b) The solution with initial condition $y(0) = 2$ is a translate to the left of the given solution.

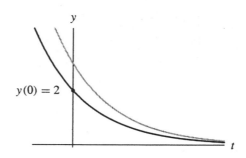

19. With $V(t) = K$, the differential equation is

$$\frac{dv_c}{dt} = \frac{K - v_c}{RC}.$$

Separating variables, we obtain

$$\frac{dv_c}{K - v_c} = \frac{dt}{RC}.$$

Integrating both sides yields

$$-\ln|K - v_c| = \frac{t}{RC} + k_1,$$

where k_1 is the constant of integration. Thus,

$$|K - v_c| = k_2 e^{-t/RC}$$

where $k_2 = e^{-k_1}$. We can eliminate the absolute values by allowing k_2 to assume either positive or negative values. We obtain the general solution

$$v_c = K + k e^{-t/RC}$$

where k (replacing $-k_2$) is the constant depending on the initial condition.

21. (a)

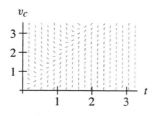

(b)

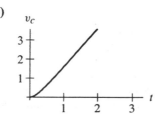

(c)

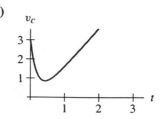

23. Our differential equation has two different expressions depending on whether $t < 1$ or $t \geq 1$. Substituting $2t$ or 2 for $V(t)$ depending on the value of t, we have

$$\frac{dv_c}{dt} = \begin{cases} (2t - v_c)/0.2 = 5(2t - v_c) & \text{for } t < 1; \\ (2 - v_c)/0.2 = 10 - 5v_c & \text{for } t \geq 1; \end{cases}$$

(a)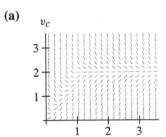

(b)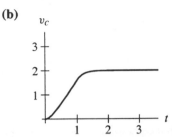

(c)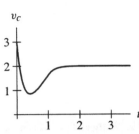

(e) For $t < 1$, the graph of the solution is identical to the one for $t < 1$ in Exercise 21. For $t \geq 1$, the graph of the solution is qualitatively similar to the one for $t \geq 1$ in Exercise 22. The complete graph is the "combination" of the graphs from Exercises 21 and 22.

EXERCISES FOR SECTION 1.4

1.

Table 1.1
Results of Euler's method

k	t_k	y_k	m_k
0	0	3	7
1	0.5	6.5	14
2	1.0	13.5	28
3	1.5	27.5	56
4	2.0	55.5	

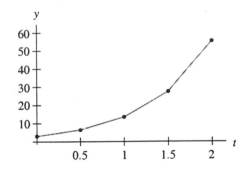

3.

Table 1.2
Results of Euler's method (to two decimal places)

k	t_k	y_k	m_k
0	0	2	0.1
1	0.5	2.5	2.25
2	1.0	3.625	6.89
3	1.5	7.07	36.84
4	2.0	25.49	

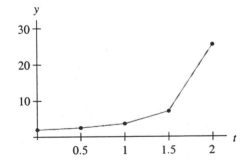

5.

Table 1.3
Results of Euler's method

k	t_k	w_k	m_k
0	0	4	−5
1	1	−1	0
2	2	−1	0
3	3	−1	0
4	4	−1	0
5	5	−1	

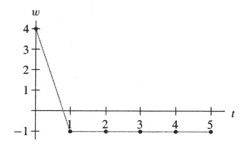

7.

Table 1.4
Results of Euler's method (shown rounded to two decimal places)

k	t_k	y_k	m_k
0	0	2	2.72
1	0.5	3.36	1.81
2	1.0	4.27	1.60
3	1.5	5.06	1.48
4	2.0	5.81	

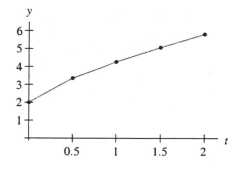

9. Because the differential equation is autonomous, the computation that determines y_{k+1} from y_k depends only on y_k and Δt and not on the actual value of t_k. Hence the approximate y-values that are obtained in both exercises are the same. It is useful to think about this fact in terms of the slope field of an autonomous equation.

11. As the solution approaches the equilibrium solution corresponding to $w = 3$, its slope decreases. We do not expect the solution to "jump over" an equilibrium solution (see the Existence and Uniqueness Theorem in Section 1.5).

13.

Table 1.5
Results of Euler's method with $\Delta t = 1.0$ (shown to two decimal places)

k	t_k	y_k	m_k
0	0	1	1
1	1	2	0
2	2	2	0
3	3	2	0
4	4	2	

Table 1.6
Results of Euler's method with $\Delta t = 0.5$ (shown to two decimal places)

k	t_k	y_k	m_k	k	t_k	y_k	m_k
0	0	1	1	5	2.5	1.97	0.02
1	0.5	1.5	0.5	6	3.0	1.98	0.02
2	1.0	1.75	0.26	7	3.5	1.99	0.02
3	1.5	1.88	0.12	8	4.0	2.0	
4	2.0	1.94	0.06				

Table 1.7
Results of Euler's method with $\Delta t = 0.25$ (shown to two decimal places)

k	t_k	y_k	m_k	k	t_k	y_k	m_k
0	0	1	1	9	2.25	1.92	0.08
1	0.25	1.25	0.76	10	2.50	1.94	0.06
2	0.50	1.44	0.56	11	2.75	1.96	0.04
3	0.75	1.58	0.40	12	3.0	1.97	0.03
4	1.0	1.68	0.32	13	3.25	1.98	0.02
5	1.25	1.76	0.24	14	3.50	1.98	0.02
6	1.50	1.82	0.18	15	3.75	1.99	0.01
7	1.75	1.87	0.13	16	4.0	1.99	
8	2.0	1.90	0.10				

From the differential equation, we see that dy/dt is positive and decreasing as long as $y(0) = 1$ and $y(t) < 2$ for $t > 0$. Therefore, $y(t)$ is increasing, and its graph is concave down. Since Euler's method uses line segments to approximate the graph of the actual solution, the approximate solutions will always be greater than the actual solution. This error decreases as the step size decreases.

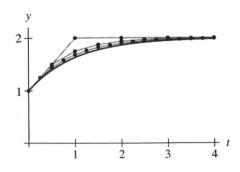

15.

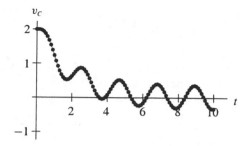

Graph of approximate solution obtained using Euler's method with $\Delta t = 0.1$.

17.

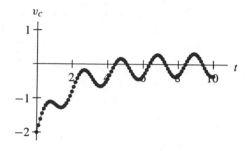

Graph of approximate solution obtained using Euler's method with $\Delta t = 0.1$.

19. (a)

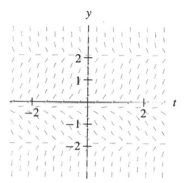

(b)

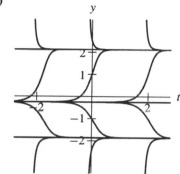

(c) The roots of $p(y)$ correspond to the equilibrium solutions.

(d) From the graphs of the solutions, we see that there are three equilibrium points, hence three real roots. We can find two of the roots by using Euler's method with a positive value of Δt. The other root can be found using Euler's method with a negative Δt. Since solutions tend toward the root, we can use a fairly large step size to obtain the required accuracy, but since this is only an approximate solution, it is best to double check the location of the root by substituting the value into $p(y)$. The approximate roots are $y \approx 2.115$, $y \approx -1.861$ and $y \approx -0.254$.

EXERCISES FOR SECTION 1.5

1. Since the constant function $y_1(t) = 3$ for all t is a solution, then the graph of any other solution $y(t)$ with $y(0) < 3$ cannot cross the line $y = 3$ by the Uniqueness Theorem. So $y(t) < 3$ for all t in the domain of $y(t)$.

3. Because $y_2(0) < y(0) < y_1(0)$, we know that

$$-t^2 = y_2(t) < y(t) < y_1(t) = t + 2$$

for all t. This restricts how large positive or negative $y(t)$ can be for a given value of t (that is, between $-t^2$ and $t+2$). As $t \to -\infty$, $y(t) \to -\infty$ between $-t^2$ and $t+2$ ($y(t) \to -\infty$ as $t \to -\infty$ at least linearly, but no faster than quadratically).

5. The Existence Theorem implies that a solution with this initial condition exists, at least for a small t-interval about $t = 0$. This differential equation has equilibrium solutions $y_1(t) = 0$, $y_2(t) = 2$, and $y_3(t) = 3$. Since $y(0) = 4$, the Uniqueness Theorem implies that $y(t) > 3$ for all t in the domain of $y(t)$. Also, $dy/dt > 0$ for all $y > 3$, so the solution $y(t)$ is increasing for all t in its domain.

7. Because $0 < y(0) < 2$ and $y_1(t) = 0$ and $y_2(t) = 2$ are equilibrium solutions of the differential equation, we know that $0 < y(t) < 2$ for all t by the Uniqueness Theorem. Also, $dy/dt > 0$ for $0 < y < 2$, so dy/dt is always positive for this solution. Hence, $y(t) \to 2$ as $t \to \infty$, and $y(t) \to 0$ as $t \to -\infty$.

9. (a) To check that $y_1(t) = t^2$ is a solution, we compute

$$\frac{dy_1}{dt} = 2t$$

and

$$-y_1^2 + y_1 + 2y_1t^2 + 2t - t^2 - t^4 = -(t^2)^2 + (t^2) + 2(t^2)t^2 + 2t - t^2 - t^4$$
$$= 2t.$$

To check that $y_2(t) = t^2 + 1$ is a solution, we compute

$$\frac{dy_2}{dt} = 2t$$

and

$$-y_2^2 + y_2 + 2y_2t^2 + 2t - t^2 - t^4 = -(t^2+1)^2 + (t^2+1) + 2(t^2+1)t^2$$
$$+ 2t - t^2 - t^4$$
$$= 2t.$$

(b) The initial values of the two solutions are $y_1(0) = 0$ and $y_2(0) = 1$. Thus if $y(t)$ is a solution and $y_1(0) = 0 < y(0) < 1 = y_2(0)$, then we can apply the Uniqueness Theorem to obtain

$$y_1(t) = t^2 < y(t) < t^2 + 1 = y_2(t)$$

for all t. Note that since the differential equation satisfies the hypothesis of the Existence and Uniqueness Theorem over the entire ty-plane, we can continue to extend the solution as long as it does not escape to $\pm\infty$ in finite time. Since it is bounded above and below by solutions that exist for all time, $y(t)$ is defined for all time also.

(c)

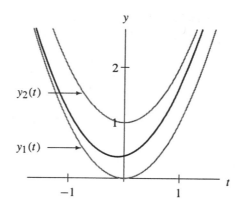

11. This exercise shows that solutions of autonomous equations cannot have local maximums or minimums. Hence they must be either constant or monotonically increasing or monotonically decreasing. A useful corollary is that a function $y(t)$ that oscillates cannot be the solution of an autonomous differential equation.

(a) Note $dy_1/dt = 0$ at $t = t_0$ because $y_1(t)$ has a local maximum. Because $y_1(t)$ is a solution, we know that $dy_1/dt = f(y_1(t))$ for all t in the domain of $y_1(t)$. In particular,

$$0 = \left.\frac{dy_1}{dt}\right|_{t=t_0} = f(y_1(t_0)) = f(y_0),$$

so $f(y_0) = 0$.

(b) This differential equation is autonomous, so the slope marks along any given horizontal line are parallel. Hence, the slope marks along the line $y = y_0$ must all have zero slope.

(c) For all t,

$$\frac{dy_2}{dt} = \frac{d(y_0)}{dt} = 0$$

because the derivative of a constant function is zero, and for all t

$$f(y_2(t)) = f(y_0) = 0.$$

So $y_2(t)$ is a solution.

(d) By the Uniqueness Theorem, we know that two solutions that are in the same place at the same time are the same solution. We have $y_1(t_0) = y_0 = y_2(t_0)$. Moreover, $y_1(t)$ is assumed to be a solution, and we showed that $y_2(t)$ is a solution in parts (a) and (b) of this exercise. So $y_1(t) = y_2(t)$ for all t. In other words, $y_1(t) = y_0$ for all t.

(e) Follow the same four steps as before. We still have $dy_1/dt = 0$ at $t = t_0$ because y_1 has a local minimum at $t = t_0$.

13. The key observation is that the differential equation is not defined when $t = 0$.

(a) Note that $dy_1/dt = 0$ and $y_1/t^2 = 0$, so $y_1(t)$ is a solution.

(b) Separating variables, we have

$$\int \frac{dy}{y} = \int \frac{dt}{t^2}.$$

Solving for y we obtain $y(t) = ce^{-1/t}$, where c is any constant. Thus, for any real number c, define the function $y_c(t)$ by

$$y_c(t) = \begin{cases} 0 & \text{for } t \le 0; \\ ce^{-1/t} & \text{for } t > 0. \end{cases}$$

For each c, $y_c(t)$ satisfies the differential equation for all $t \ne 0$.

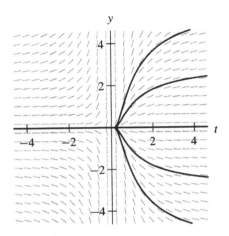

There are infinitely many solutions of the
form $y_c(t)$ that agree with $y_1(t)$ for $t < 0$.

(c) Note that $f(t, y) = y/t^2$ is not defined at $t = 0$. Therefore, we *cannot* apply the Uniqueness Theorem for the initial condition $y(0) = 0$. The "solution" $y_c(t)$ given in part (b) actually represents two solutions, one for $t < 0$ and one for $t > 0$.

15. (a) The equation is separable, so we obtain

$$\int (y+1)\, dy = \int \frac{dt}{t-2}.$$

Solving for y with help from the quadratic formula yields the general solution

$$y(t) = -1 \pm \sqrt{1 + \ln(c(t-2)^2)}$$

where c is a constant. Substituting the initial condition $y(0) = 0$ and solving for c, we have

$$0 = -1 \pm \sqrt{1 + \ln(4c)},$$

and thus $c = 1/4$. The desired solution is therefore

$$y(t) = -1 + \sqrt{1 + \ln((1 - t/2)^2)}$$

(b) The solution is defined only when $1 + \ln((1 - t/2)^2) \ge 0$, that is, when $|t - 2| \ge 2/\sqrt{e}$. Therefore, the domain of the solution is

$$t \le 2(1 - 1/\sqrt{e}).$$

(c) As $t \to 2(1 - 1/\sqrt{e})$, then $1 + \ln((1 - t/2)^2) \to 0$. Thus

$$\lim_{t \to 2(1-1/\sqrt{e})} y(t) = -1.$$

Note that the differential equation is not defined at $y = -1$. Also, note that

$$\lim_{t \to -\infty} y(t) = \infty.$$

17. **(a)** The equation is separable. Separating variables we obtain

$$\int (y - 2)\, dy = \int t\, dt.$$

Solving for y with help from the quadratic formula yields the general solution

$$y(t) = 2 \pm \sqrt{t^2 + c}.$$

To find c, we let $t = -1$ and $y = 0$, and we obtain $c = 3$. The desired solution is therefore $y(t) = 2 - \sqrt{t^2 + 3}$

(b) Since $t^2 + 2$ is always positive and $y(t) < 2$ for all t, the solution $y(t)$ is defined for all real numbers.

(c) As $t \to \pm\infty$, $t^2 + 3 \to \infty$. Therefore,

$$\lim_{t \to \pm\infty} y(t) = -\infty.$$

EXERCISES FOR SECTION 1.6

1. The equilibrium points of $dy/dt = f(y)$ are the numbers y where $f(y) = 0$. For

$$f(y) = 3y(1 - y),$$

the equilibrium points are $y = 0$ and $y = 1$. Since $f(y)$ is negative for $y < 0$, positive for $0 < y < 1$, and negative for $y > 1$, the equilibrium point $y = 0$ is a source and the equilibrium point $y = 1$ is a sink.

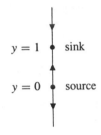

$y = 1$ • sink

$y = 0$ • source

3. The equilibrium points of $dy/dt = f(y)$ are the numbers y where $f(y) = 0$. For $f(y) = \cos y$, the equilibrium points are $y = \pi/2 + n\pi$, where $n = 0, \pm 1, \pm 2, \dots$. Since $\cos y > 0$ for $-\pi/2 < y < \pi/2$ and $\cos y < 0$ for $\pi/2 < y < 3\pi/2$, we see that the equilibrium point at $y = \pi/2$ is a sink. Since the sign of $\cos y$ alternates between positive and negative in a period fashion, we see that the equilibrium points at $y = \pi/2 + 2n\pi$ are sinks and the equilibrium points at $y = 3\pi/2 + 2n\pi$ are sources.

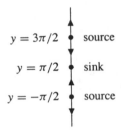

$y = 3\pi/2$ • source

$y = \pi/2$ • sink

$y = -\pi/2$ • source

5. The equilibrium points of $dw/dt = f(w)$ are the numbers w where $f(w) = 0$. For $f(w) = (w-2)\sin w$, the equilibrium points are $w = 2$ and $w = n\pi$, where $n = 0, \pm 1, \pm 2, \dots$. The sign of $(w - 2)\sin w$ alternates between positive and negative at successive zeros. It is positive for $-\pi < w < 0$ and negative for $0 < w < 2$. Therefore, $w = 0$ is a sink, and the equilibrium points alternate back and forth between sources and sinks.

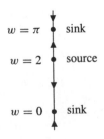

$w = \pi$ • sink

$w = 2$ • source

$w = 0$ • sink

7. The derivative dw/dt is always positive, so there are no equilibrium points, and all solutions increase for all time.

9. The equilibrium points of $dy/dt = f(y)$ are the numbers y where $f(y) = 0$. For $f(y) = -1 + \cos y$, the equilibrium points are $y = 2n\pi$, where $n = 0, \pm1, \pm2, \dots$. Since $f(y)$ is nonpositive for all values of y, all of the equilibrium points are nodes.

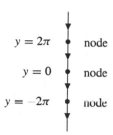

11. The equilibrium points of $dy/dt = f(y)$ are the numbers y where $f(y) = 0$. For $f(y) = y \ln |y|$, there are equilibrium points at $y = \pm1$. In addition, although the function $f(y)$ is technically undefined at $y = 0$, the limit of $f(y)$ as $y \to 0$ is 0. Thus we can treat $y = 0$ as another equilibrium point. Since $f(y) < 0$ for $y < -1$ and $0 < y < 1$, and $f(y) > 0$ for $y > 1$ and $-1 < y < 0$, $y = -1$ is a source, $y = 0$ is a sink, and $y = 1$ is a source.

13.

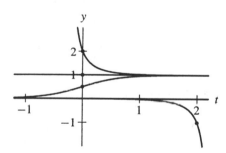

15.

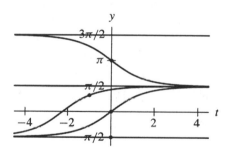

17.

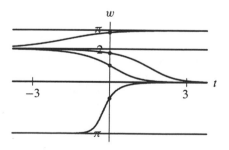

19.

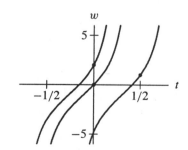

21.

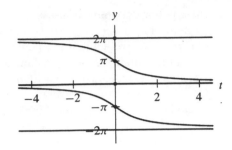

23. The initial value $y(0) = 1$ is between the equilibrium points $y = 2 - \sqrt{2}$ and $y = 2 + \sqrt{2}$. Also, $dy/dt < 0$ for $2 - \sqrt{2} < y < 2 + \sqrt{2}$. Hence the solution is decreasing and tends toward the smaller equilibrium point $y = 2 - \sqrt{2}$ as $t \to \infty$. It tends toward the larger equilibrium point $y = 2 + \sqrt{2}$ as $t \to -\infty$.

25. The initial value $y(0) = -10$ is below both of the equilibrium points. Since dy/dt is positive for $y < 2 - \sqrt{2}$, the solution is increasing for all t and tends to the equilibrium point $y = 2 - \sqrt{2}$ as $t \to \infty$. As t decreases, it becomes unbounded in the negative direction in finite time.

27. The initial value $y(3) = 1$ is between the equilibrium points $y = 2 - \sqrt{2}$ and $y = 2 + \sqrt{2}$. Also, $dy/dt < 0$ for $2 - \sqrt{2} < y < 2 + \sqrt{2}$. Hence the solution is decreasing and tends toward the smaller equilibrium point $y = 2 - \sqrt{2}$ as $t \to \infty$. It tends toward the larger equilibrium point $y = 2 + \sqrt{2}$ as $t \to -\infty$.

29. The function $f(y)$ has two zeros $\pm y_0$, where y_0 is some positive number. So the differential equation $dy/dt = f(y)$ has two equilibrium solutions, one for each zero. Also, $f(y) < 0$ if $-y_0 < y < y_0$ and $f(y) > 0$ if $y < -y_0$ or if $y > y_0$. Hence y_0 is a source and $-y_0$ is a sink.

31. The function $f(y)$ has two zeros, one positive and one negative. We denote them as y_1 and y_2, where $y_1 < y_2$. So the differential equation $dy/dt = f(y)$ has two equilibrium solutions, one for each zero. Also, $f(y) > 0$ if $y_1 < y < y_2$ and $f(y) < 0$ if $y < y_1$ or if $y > y_2$. Hence y_1 is a source and y_2 is a sink.

33. Since there are two equilibrium points, the graph of $f(y)$ must touch the y-axis at two distinct numbers y_1 and y_2. Assume that $y_1 < y_2$. Since the arrows point up if $y < y_1$ and if $y > y_2$, we must have $f(y) > 0$ for $y < y_1$ and for $y > y_2$. Similarly, $f(y) < 0$ for $y_1 < y < y_2$.

The precise location of the equilibrium points is not given, and the direction of the arrows on the phase line is determined only by the sign (and not the magnitude) of $f(y)$. So the following graph is one of many possible answers.

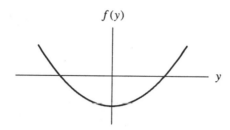

35. Since there are four equilibrium points, the graph of $f(y)$ must touch the y-axis at four distinct numbers y_1, y_2, y_3, and y_4. We assume that $y_1 < y_2 < y_3 < y_4$. Since the arrows point up only if $y_1 < y < y_2$ or if $y_2 < y < y_3$, we must have $f(y) > 0$ for $y_1 < y < y_2$ and for $y_2 < y < y_3$. Moreover, $f(y) < 0$ if $y < y_1$, if $y_3 < y < y_4$, or if $y > y_4$. Therefore, the graph of f crosses the y-axis at y_1 and y_3, but it is tangent to the y-axis at y_2 and y_4.

The precise location of the equilibrium points is not given, and the direction of the arrows on the phase line is determined only by the sign (and not the magnitude) of $f(y)$. So the following graph is one of many possible answers.

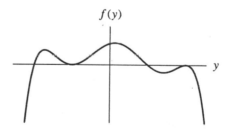

37. **(a)** In terms of the phase line with $P \geq 0$, there are three equilibrium points. If we assume that $f(P)$ is differentiable, then a decreasing population at $P = 100$ implies that $f(P) < 0$ for $P > 50$. An increasing population at $P = 25$ implies that $f(P) > 0$ for $10 < P < 50$. These assumptions leave two possible phase lines since the arrow between $P = 0$ and $P = 10$ is undetermined.

$P = 50$

$P = 10$
$P = 0$

(b) Given the observations in part (a), we see that there are two basic types of graphs that go with the assumptions. However, there are many graphs that correspond to each possibility. The following two graphs are representative.

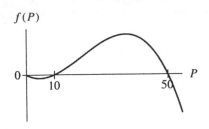
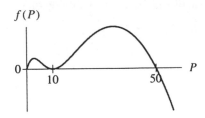

(c) The functions $f(P) = P(P - 10)(50 - P)$ and $f(P) = P(P - 10)^2(50 - P)$ respectively are two examples but there are many others.

39. The equilibrium points occur at solutions of $dy/dt = y^2 + a = 0$. For $a > 0$, there are no equilibrium points. For $a = 0$, there is one equilibrium point, $y = 0$. For $a < 0$, there are two equilibrium points, $y = \pm\sqrt{-a}$.

To draw the phase lines, note that:

- If $a > 0$, $dy/dt = y^2 + a > 0$, so the solutions are always increasing.
- If $a = 0$, $dy/dt > 0$ unless $y = 0$. Thus, $y = 0$ is a node.
- For $a < 0$, $dy/dt < 0$ for $-\sqrt{-a} < y < \sqrt{-a}$, and $dy/dt > 0$ for $y < -\sqrt{-a}$ and for $y > \sqrt{-a}$.

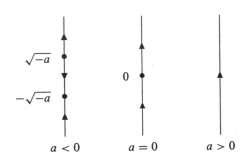

(a) The phase lines for $a < 0$ are qualitatively the same, and the phase lines for $a > 0$ are qualitatively the same.

(b) The phase line undergoes a qualitative change at $a = 0$.

41. **(a)** Because $f(y)$ is continuous we can use the Intermediate Value Theorem to say that there must be a zero of $f(y)$ between -10 and 10. This value of y is an equilibrium point of the differential equation. In fact, $f(y)$ must cross from positive to negative, so if there is a single equilibrium point, it must be a sink (see part (b)).

(b) We know that $f(y)$ must cross the y-axis between -10 and 10. Moreover, it must cross from positive to negative because $f(-10)$ is positive and $f(10)$ is negative. Where $f(y)$ crosses the

y-axis from positive to negative, we have a sink. If $y = 1$ is a source, then crosses the y-axis from negative to positive at $y = 1$. Hence, $f(y)$ must cross the y-axis from positive to negative at least once between $y = -10$ and $y = 1$ and at least once between $y = 1$ and $y = 10$. There must be at least one sink in each of these intervals. (We need the assumption that the number of equilibrium points is finite to prevent cases where $f(y) = 0$ along an entire interval.)

43. (a) Because the first and second derivative are zero at y_0 and the third derivative is positive, Taylor's Theorem implies that the function $f(y)$ is approximately equal to

$$\frac{f'''(y_0)}{3!}(y - y_0)^3$$

for y near y_0. Since $f'''(y_0) > 0, f(y)$ is increasing near y_0. Hence, y_0 is a source.

(b) Just as in part (a), we see that $f(y)$ is decreasing near y_0, so y_0 is a sink.

(c) In this case, we can approximate $f(y)$ near y_0 by

$$\frac{f''(y_0)}{2!}(y - y_0)^2.$$

Since the second derivative of $f(y)$ at y_0 is assumed to be positive, $f(y)$ is positive on both sides of y_0 for y near y_0. Hence y_0 is a node.

45. One assumption of the model is that, if no people are present, then the time between trains decreases at a constant rate. Hence the term $-\alpha$ represents this assumption. The parameter α should be positive, so that $-\alpha$ makes a negative contribution to dx/dt.

The term βx represents the effect of the passengers. The parameter β should be positive so that βx contributes positively to dx/dt.

47. Note that the only equilibrium point is a source. If the initial gap between trains is too large, then x will increase without bound. If it is too small, x will decrease to zero. When $x = 0$, the two trains are next to each other, and they will stay together since $x < 0$ is not physically possible in this problem.

If the time between trains is exactly the equilibrium value $(x = \alpha/\beta)$, then theoretically $x(t)$ is constant. However, any disruption to x causes the solution to tend away from the source. Since it is very likely that some stops will have fewer than the expected number of passengers and some stops will have more, it is unlikely that the time between trains will remain constant for long.

EXERCISES FOR SECTION 1.7

1. The equilibrium points occur at solutions of $dy/dt = y^2 + a = 0$. For $a > 0$, there are no equilibrium points. For $a = 0$, there is one equilibrium point, $y = 0$. For $a < 0$, there are two equilibrium points, $y = \pm\sqrt{-a}$. Thus, $a = 0$ is a bifurcation value
To draw the phase lines, note that:

- If $a > 0, dy/dt = y^2 + a > 0$, so the solutions are always increasing.
- If $a = 0, dy/dt > 0$ unless $y = 0$. Thus, $y = 0$ is a node.

- For $a < 0$, $dy/dt < 0$ for $-\sqrt{-a} < y < \sqrt{-a}$, and $dy/dt > 0$ for $y < -\sqrt{-a}$ and for $y > \sqrt{-a}$.

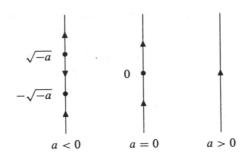

Phase lines for $a < 0$, $a = 0$, and $a > 0$.

3. The equilibrium points occur at solutions of $dy/dt = y^2 - ay + 1 = 0$. From the quadratic formula, we have

$$y = \frac{a \pm \sqrt{a^2 - 4}}{2}.$$

If $-2 < a < 2$, then $a^2 - 4 < 0$, and there are no equilibrium points. If $a > 2$ or $a < -2$, there are two equilibrium points. For $a = \pm 2$, there is one equilibrium point at $y = a/2$. The bifurcations occur at $a = \pm 2$.

To draw the phase lines, note that:

- For $-2 < a < 2$, $dy/dt = y^2 - ay + 1 > 0$, so the solutions are always increasing.
- For $a = 2$, $dy/dt = (y - 1)^2 \geq 0$, and $y = 1$ is a node.
- For $a = -2$, $dy/dt = (y + 1)^2 \geq 0$, and $y = -1$ is a node.
- For $a < -2$ or $a > 2$, let

$$y_1 = \frac{a - \sqrt{a^2 - 4}}{2} \quad \text{and} \quad y_2 = \frac{a + \sqrt{a^2 - 4}}{2}.$$

Then $dy/dt < 0$ if $y_1 < y < y_2$, and $dy/dt > 0$ if $y < y_1$ or $y > y_2$.

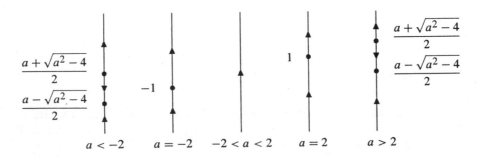

The five possible phase lines.

5. The equilibrium points occur at solutions of

$$\frac{dy}{dt} = y^6 - 2y^3 + \alpha = 0.$$

Using the quadratic equation to solve for y^3, we obtain

$$y^3 = \frac{2 \pm \sqrt{4 - 4\alpha}}{2}.$$

So the equilibrium points are at

$$y = \left(1 \pm \sqrt{1 - \alpha}\right)^{1/3}.$$

If $\alpha > 1$, there are no equilibrium points because this equation has no real solutions. If $\alpha < 1$, the differential equation has two equilibrium points. A bifurcation occurs at $\alpha = 1$ where the differential equation has one equilibrium point at $y = 1$.

7. **(a)** For all $C \geq 0$, the equation has a source at $P = C/k$, and this is the only equilibrium point. Hence all of the phase lines are qualitatively the same, and there are no bifurcation values for C.

(b) If $P(0) > C/k$, the corresponding solution $P(t) \to \infty$ at an exponential rate as $t \to \infty$, and if $P(0) < C/k$, $P(t) \to -\infty$, passing through "extinction" ($P = 0$) after a finite time.

9. **(a)** A model of the fish population that includes fishing is

$$\frac{dP}{dt} = 2P - \frac{P^2}{50} - 3L,$$

where L is the number of licenses issued. The coefficient of 3 represents the average catch of 3 fish per year. As L is increased, the two equilibrium points for $L = 0$ (at $P = 0$ and $P = 100$) will move together. If L is sufficiently large, there are no equilibrium points. Hence we wish to pick L as large as possible so that there is still an equilibrium point present. In other words, we want the bifurcation value of L. The bifurcation value of L occurs if the equation

$$\frac{dP}{dt} = 2P - \frac{P^2}{50} - 3L = 0$$

has just one solution for P in terms of L. Using the quadratic formula, we see that there is exactly one equilibrium point if $L = 50/3$. Since this value of L is not an integer, the largest number of licenses that should be allowed is 16.

(b) If we allow the fish population to come to equilibrium then the population will be at the carrying capacity, which is $P = 100$ if $L = 0$. If we then allow 16 licenses to be issued, we expect that the population is a solution to the new model with $L = 16$ and initial population $P = 100$. The model becomes

$$\frac{dP}{dt} = 2P - \frac{P^2}{50} - 48,$$

which has a source at $P = 40$ and a sink at $P = 60$.

Thus, any initial population greater than 40 when fishing begins tends to the equilibrium level $P = 60$. If the initial population of fish was less than 40 when fishing begins, then the model predicts that the population will decrease to zero in a finite amount of time.

(c) The maximum "number" of licenses is $16\frac{2}{3}$. With $L = 16\frac{2}{3}$, there is an equilibrium at $P = 50$. This equilibrium is a node, and if $P(0) > 50$, the population will approach 50 as t increases. However, it is dangerous to allow this many licenses since an unforeseen event might cause the death of a few extra fish. That event would push the number of fish below the equilibrium value of $P = 50$. In this case, $dP/dt < 0$, and the population decreases to extinction.

If, however, we restrict to $L = 16$ licenses, then there are two equilibria, a sink at $P = 60$ and source at $P = 40$. As long as $P(0) > 40$, the population will tend to 60 as t increases. In this case, we have a small margin of safety. If $P \approx 60$, then it would have to drop to less than 40 before the fish are in danger of extinction.

11. (a)

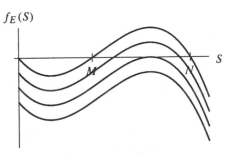

(b) As E increases, the graph of $f_E(S)$ is translated down until it has a double root between M and N. This value of E, which we denote by E_b, is the bifurcation value for this model. If $0 < E < E_b$, $f_E(S)$ has three real roots S_1, S_2, and S_3, where $S_1 < 0$ and $M < S_2 < S_3 < N$. Note that $f_E(S) > 0$ if $S_2 < S < S_3$ and $f_E(S) < 0$ if $S_1 < S < S_2$ or $S > S_3$. Thus, S_2 is a source and S_3 is a sink. Consequently, if the fox squirrel population starts above S_2, it tends to the carrying capacity S_3.

If $E > E_b$, $f_E(S)$ no longer has any positive roots. Hence, there are no positive equilibria. Moreover, $f_E(S) < 0$ for all $S > 0$, and thus the fox squirrel population will plummet toward extinction.

(c) We must find the value of E for which $f_E(S)$ has a double root. Differentiation with respect to S yields

$$f'(S) = -\frac{3k}{MN}\left(S^2 - \frac{2}{3}(M + N)S + \frac{MN}{3}\right).$$

Setting $f'(S) = 0$ gives the two critical points

$$S = \frac{M + N \pm \sqrt{M^2 + N^2 - MN}}{3}.$$

Let S_c denote the largest of these two numbers.

As E increases, the graph of $f_E(S)$ shifts down. To find the bifurcation value E_b, we need to find the value of E such that $f_E(S_c) = 0$. Therefore, we set

$$f_E\left(\frac{M + N + \sqrt{M^2 + N^2 - MN}}{3}\right) = 0$$

and solve for E. After much computation, we find that

$$E_b = \frac{k}{27MN}\left((N - 2M)(M + N)(2N - M) + 2(M^2 - MN + N^2)^{3/2}\right).$$

13. **(a)** (i) The harvesting rate is constant.

(ii) The harvesting rate decreases linearly in P. Expressed as a relative rate (that is, as a ratio of the rate divided by the population), it is $1/3 \approx 33\%$. In other words, as a percentage of the population, the harvesting rate is 33%.

(iii) The harvesting rate is $P/(1+2P)$, which decreases as P decreases. In this case, the relative harvesting rate is $1/(1+2P)$. For $P \approx 1$, the relative harvesting rate is approximately 33%, and for $P \approx 0$, the relative harvesting rate is approximately 100%.

(iv) The harvesting rate is $3P/(2+7P)$, which decreases as P decreases. In this case, the relative harvesting rate is $3/(2+7P)$. For $P \approx 1$, the relative harvesting rate is approximately 33%, and for $P \approx 0$, the relative harvesting rate is approximately 150%.

(b) (i) For this model, dP/dt is always negative, so there are no equilibria. Hence, the population becomes extinct.

(ii) This model has an equilibrium point at $P = 2/3$, and for $P > 2/3, dP/dt < 0$. Hence, this solution tends toward the equilibrium at $P = 2/3$.

(iii) This model has an equilibrium point at $P = 1/2$, and for $P > 1/2, dP/dt < 0$. Hence, this solution tends toward the equilibrium at $P = 1/2$.

(iv) This model has an equilibrium point at $P = 0$, and for $P > 0, dP/dt < 0$. Hence, this solution tends toward the equilibrium at $P = 0$.

(c) In model (i), the population goes extinct (relatively quickly). Models (ii) and (iii) predict a new equilibrium level of the population. The equilibrium population for (ii) is larger than the equilibrium for (iii). In model (iv), the population approaches extinction but never actually reaches extinction. With the aid of HPGSolver, we see that the population decreases rapidly between $t = 0$ and $t = 10$. Then it almost levels out between $t = 10$ and $t = 30$ before it again falls quickly becoming asymptotic to zero by $t = 50$.

15. **(a)** If $a = 0$, there is a single equilibrium point at $y = 0$. For $a \neq 0$, the equilibrium points occur at $y = 0$ and $y = a$. If $a < 0$, the equilibrium point at $y = 0$ is a sink and the equilibrium point at $y = a$ is a source. If $a > 0$, the equilibrium point at $y = 0$ is a source and the equilibrium point at $y = a$ is a sink.

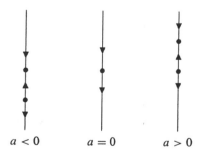

$a < 0$ $a = 0$ $a > 0$

Phase lines for $dy/dt = ay - y^2$.

(b) Given the results in part (a), there is one bifurcation value, $a = 0$.

(c) The equilibrium points satisfy the equation

$$r + ay - y^2 = 0.$$

Solving it, we obtain

$$y = \frac{a \pm \sqrt{a^2 + 4r}}{2}.$$

Hence, there are no equilibrium points if $a^2 + 4r < 0$, one equilibrium point if $a^2 + 4r = 0$, and two equilibrium points if $a^2 + 4r > 0$.

If $r > 0$, we always have two equilibrium points.

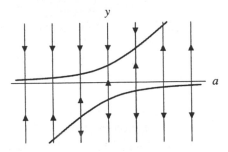

The bifurcation diagram for $r > 0$.

(d) If $r < 0$, there are no equilibrium points if $a^2 + 4r < 0$. In other words, there are no equilibrium points if $-2\sqrt{-r} < a < 2\sqrt{-r}$. If $a = \pm 2\sqrt{-r}$, there is a single equilibrium point, and if $|a| > 2\sqrt{-r}$, there are two equilibrium points.

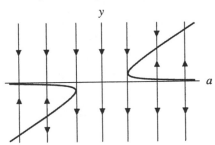

The bifurcation diagram for $r < 0$.

17. If the index is -1 for $\alpha = 0$, then $f_0(0) > 0$ and $f_0(1) < 0$ because there must be one more sink than source between $y = 0$ and $y = 1$ for $dy/dt = f_0(y)$. Since the points $y = 0$ and $y = 1$ are never equilibrium points for any α, the signs of $f_\alpha(0)$ and $f_\alpha(1)$ cannot change as α varies. Thus, $f_\alpha(0) > 0$ and $f_\alpha(1) < 0$ for all α. Hence, the graph of $f_\alpha(y)$ must cross the y-axis from positive to negative at least once. Consequently, $dy/dt = f_\alpha(y)$ has at least one sink in $0 \le y \le 1$ for all α.

19. Since $f_0(0) > 0$ and the index is zero in $0 \le y \le 1$ for $\alpha = 0$, $f_0(1) > 0$. Since $f_\alpha(1) \ne 0$ for $0 \le \alpha \le 1$, $f_\alpha(1) > 0$ for $0 \le \alpha \le 1$. Hence, the index is zero in $0 \le y \le 1$ for $0 \le \alpha \le 1$.

Since $f_\alpha(y)$ is continuous in α and y, it must have a minimum value m on the rectangle determined by $0 \le \alpha \le 1$ and $0 \le y \le 1$ in the αy-plane. If $m \ge 0$, M can be any positive number. If $m < 0$, we choose M to be any number greater than $|m|$.

We have $f_\alpha(0) + r > 0$ and $f_\alpha(1) + r > 0$ for $0 \le r \le M$, so the index of $f_\alpha(y) + r$ must be zero for $0 \le r \le M$.

21. We have

$$\frac{dy}{dt} = y^4 + \alpha y^2 = y^2(y^2 + \alpha).$$

If $\alpha > 0$, there is one equilibrium point at $y = 0$, and $dy/dt > 0$ otherwise. Hence $y = 0$ is a node.

If $\alpha < 0$, $dy/dt = 0$ at $y = 0$ and $y = \pm\sqrt{-\alpha}$. From the sign of $y^4 + \alpha y^2$, we know that $y = 0$ is a node, $y = -\sqrt{-\alpha}$ is a sink, and $y = \sqrt{-\alpha}$ is a source.

23. For $\alpha < 0$, both $y^2 + \alpha$ and $y^4 + \alpha$ have two roots. These four roots are distinct. Therefore, the product, $y(y + \alpha)(y^2 + \alpha)(y^4 + \alpha)$ has six distinct roots for $\alpha < 0$. For $\alpha = 0$, the differential equation is

$$\frac{dy}{dt} = y^8,$$

which has only one equilibrium point, $y = 0$.

EXERCISES FOR SECTION 1.8

1. We rewrite the equation in the form

$$\frac{dy}{dt} + \frac{y}{t} = 2$$

and note that the integrating factor is

$$\mu(t) = e^{\int (1/t)\, dt} = e^{\ln t} = t.$$

Multiplying both sides by $\mu(t)$, we obtain

$$t\frac{dy}{dt} + y = 2t.$$

Applying the Product Rule to the left-hand side, we see that this equation is the same as

$$\frac{d(ty)}{dt} = 2t,$$

and integrating both sides with respect to t, we obtain

$$ty = t^2 + c,$$

where c is an arbitrary constant. The general solution is

$$y(t) = \frac{1}{t}(t^2 + c) = t + \frac{c}{t}.$$

3. We rewrite the equation in the form

$$\frac{dy}{dt} - y = -3e^{-t}$$

and note that the integrating factor is

$$\mu(t) = e^{\int -1\,dt} = e^{-t}.$$

Multiplying both sides by $\mu(t)$, we obtain

$$e^{-t}\frac{dy}{dt} - e^{-t}y = -3e^{-2t}.$$

Applying the Product Rule to the left-hand side, we see that this equation is the same as

$$\frac{d(e^{-t}y)}{dt} = -3e^{-2t}.$$

and integrating both sides with respect to t, we obtain

$$e^{-t}y = \tfrac{3}{2}e^{-2t} + c$$

where c is an arbitrary constant. The general solution is

$$y(t) = \tfrac{3}{2}e^{-t} + ce^{t}.$$

5. We rewrite the equation in the form

$$\frac{dy}{dt} - \frac{2t}{1+t^2}\,y = \frac{2}{1+t^2}$$

and note that the integrating factor is

$$\mu(t) = e^{\int(-2t/(1+t^2))\,dt} = e^{-\ln(1+t^2)} = \left(e^{\ln(1+t^2)}\right)^{-1} = \frac{1}{1+t^2}.$$

Multiplying both sides by $\mu(t)$, we obtain

$$\frac{1}{1+t^2}\frac{dy}{dt} - \frac{2t}{(1+t^2)^2}\,y = \frac{2}{(1+t^2)^2}.$$

Applying the Product Rule to the left-hand side, we see that this equation is the same as

$$\frac{d}{dt}\left(\frac{y}{1+t^2}\right) = \frac{2}{(1+t^2)^2}.$$

An antiderivative of $2/(1+t^2)^2$ is

$$\arctan t + \frac{t}{1+t^2},$$

so integrating both sides with respect to t, we obtain

$$\frac{y}{1+t^2} = \arctan t + \frac{t}{1+t^2} + c,$$

where c is an arbitrary constant. The general solution is

$$y(t) = t + (1+t^2)(\arctan(t) + c).$$

7. We rewrite the equation in the form

$$\frac{dy}{dt} + y = t^2$$

and note that the integrating factor is

$$\mu(t) = e^{\int 1\, dt} = e^t.$$

Multiplying both sides by $\mu(t)$, we obtain

$$e^t \frac{dy}{dt} + e^t y = t^2 e^t.$$

Applying the Product Rule to the left-hand side, we see that this equation is the same as

$$\frac{d(e^t y)}{dt} = t^2 e^t,$$

and integrating both sides with respect to t, we obtain

$$e^t y = \int t^2 e^t \, dt.$$

The integral on the right-hand side can be computed by integrating by parts twice with $dv = e^t dt$ each time. We have

$$e^t y = (t^2 - 2t + 2)e^t + c,$$

where c is an arbitrary constant. The general solution is

$$y(t) = t^2 - 2t + 2 + ce^{-t}.$$

9. We rewrite the equation in the form

$$\frac{dy}{dt} + \frac{y}{1+t} = 2$$

and note that the integrating factor is

$$\mu(t) = e^{\int (1/(1+t))\, dt} = e^{\ln(1+t)} = 1 + t.$$

Multiplying both sides by $\mu(t)$, we obtain

$$(1+t)\frac{dy}{dt} + y = 2(1+t).$$

Applying the Product Rule to the left-hand side, we see that this equation is the same as

$$\frac{d((1+t)y)}{dt} = 2(1+t),$$

and integrating both sides with respect to t, we obtain

$$(1+t)y = 2t + t^2 + c,$$

where c is an arbitrary constant. The general solution is

$$y(t) = \frac{t^2 + 2t + c}{1 + t}.$$

To find the solution that satisfies the initial condition $y(0) = 3$, we evaluate the general solution at $t = 0$ and obtain

$$c = 3.$$

The desired solution is

$$y(t) = \frac{t^2 + 2t + 3}{1 + t}.$$

11. We rewrite the equation in the form

$$\frac{dy}{dt} + \frac{y}{t} = 2$$

and note that the integrating factor is

$$\mu(t) = e^{\int (1/t)\, dt} = e^{\ln t} = t.$$

Multiplying both sides by $\mu(t)$, we obtain

$$t\frac{dy}{dt} + y = 2t.$$

Applying the Product Rule to the left-hand side, we see that this equation is the same as

$$\frac{d(ty)}{dt} = 2t,$$

and integrating both sides with respect to t, we obtain

$$ty = t^2 + c,$$

where c is an arbitrary constant. The general solution is

$$y(t) = t + \frac{c}{t}.$$

To find the solution that satisfies the initial condition $y(1) = 3$, we evaluate the general solution at $t = 1$ and obtain

$$1 + c = 3.$$

Hence, $c = 2$, and the desired solution is

$$y(t) = t + \frac{2}{t}.$$

13. We rewrite the equation in the form

$$\frac{dy}{dt} - \frac{2y}{t} = 2t^2$$

and note that the integrating factor is

$$\mu(t) = e^{\int -(2/t)\,dt} = \frac{1}{t^2}.$$

Multiplying both sides by $\mu(t)$, we obtain

$$\frac{1}{t^2}\frac{dy}{dt} - \frac{2y}{t^3} = 2.$$

Applying the Product Rule to the left-hand side, we see that this equation is the same as

$$\frac{d}{dt}\left(\frac{y}{t^2}\right) = 2,$$

and integrating both sides with respect to t, we obtain

$$\frac{y}{t^2} = 2t + c,$$

where c is an arbitrary constant. The general solution is

$$y(t) = 2t^3 + ct^2.$$

To find the solution that satisfies the initial condition $y(-2) = 4$, we evaluate the general solution at $t = -2$ and obtain

$$-16 + 4c = 4.$$

Hence, $c = 5$, and the desired solution is

$$y(t) = 2t^3 + 5t^2.$$

15. We rewrite the equation in the form

$$\frac{dy}{dt} - (\sin t)y = 4$$

and note that the integrating factor is

$$\mu(t) = e^{\int(-\sin t)\,dt} = e^{\cos t}.$$

Multiplying both sides by $\mu(t)$, we obtain

$$e^{\cos t}\frac{dy}{dt} - e^{\cos t}(\sin t)y = 4e^{\cos t}.$$

Applying the Product Rule to the left-hand side, we see that this equation is the same as

$$\frac{d(e^{\cos t}y)}{dt} = 4e^{\cos t},$$

and integrating both sides with respect to t, we obtain

$$e^{\cos t} y = \int 4 e^{\cos t} \, dt.$$

Since the integral on the right is impossible to express using elementary functions, we write the general solution as

$$y(t) = 4 e^{-\cos t} \int e^{\cos t} \, dt.$$

17. We rewrite the equation in the form

$$\frac{dy}{dt} - \frac{y}{t^2} = 4 \cos t$$

and note that the integrating factor is

$$\mu(t) = e^{\int (-1/t^2) \, dt} = e^{1/t}.$$

Multiplying both sides by $\mu(t)$, we obtain

$$e^{1/t} \frac{dy}{dt} - \frac{e^{1/t}}{t^2} y = 4 e^{1/t} \cos t.$$

Applying the Product Rule to the left-hand side, we see that this equation is the same as

$$\frac{d(e^{1/t} y)}{dt} = 4 e^{1/t} \cos t,$$

and integrating both sides with respect to t, we obtain

$$e^{1/t} y = \int 4 e^{1/t} \cos t \, dt.$$

Since the integral on the right is impossible to express using elementary functions, we write the general solution as

$$y(t) = 4 e^{-1/t} \int e^{1/t} \cos t \, dt.$$

19. We rewrite the equation in the form

$$\frac{dy}{dt} + e^{-t^2} y = \cos t$$

and note that the integrating factor is

$$\mu(t) = e^{\int e^{-t^2} \, dt}.$$

This integral is impossible to express in terms of elementary functions. Multiplying both sides by $\mu(t)$, we obtain

$$e^{\int e^{-t^2} \, dt} \frac{dy}{dt} + e^{\int e^{-t^2} \, dt} e^{-t^2} y = e^{\int e^{-t^2} \, dt} \cos t.$$

Applying the Product Rule to the left-hand side, we see that this equation is the same as

$$\frac{d(e^{\int e^{-t^2} dt} y)}{dt} = e^{\int e^{-t^2} dt} \cos t,$$

and integrating both sides with respect to t, we obtain

$$e^{\int e^{-t^2} dt} y = \int e^{\int e^{-t^2} dt} \cos t \, dt.$$

These integrals are also impossible to compute, so we write the general solution in the form

$$y(t) = e^{-\int e^{-t^2} dt} \int e^{\int e^{-t^2} dt} \cos t \, dt.$$

21. We rewrite the equation in the form

$$\frac{dy}{dt} - aty = 4e^{-t^2}$$

and note that the integrating factor is

$$\mu(t) = e^{\int (-at) dt} = e^{-at^2/2}.$$

Multiplying both sides by $\mu(t)$, we obtain

$$e^{-at^2/2} \frac{dy}{dt} - ate^{-at^2/2} y = 4e^{-t^2} e^{-at^2/2}.$$

Applying the Product Rule to the left-hand side and simplifying the right-hand side, we see that this equation is the same as

$$\frac{d(e^{-at^2/2} y)}{dt} = 4e^{-(1+a/2)t^2}.$$

Integrating both sides with respect to t, we obtain

$$e^{-at^2/2} y = \int 4e^{-(1+a/2)t^2} dt.$$

The integral on the right-hand side can be expressed in terms of elementary functions only if $1 + a/2 = 0$ (that is, if the factor involving e^{t^2} really isn't there). Hence, the only value of a that yields an integral we can express in closed form is $a = -2$ (see Exercise 4).

23. **(a)** The differential equation modeling the problem is

$$\frac{dP}{dt} = .04P + 520$$

where 520 is the amount of money added to the account per year. (Assuming 52 weeks in a year and a "continuous deposit".)

(b) Rewriting the equation as $dP/dt - .04P = 520$, we see that the integrating factor is

$$\mu(t) = e^{\int -.04\,dt} = e^{-.04t}.$$

Multiplying both sides by $\mu(t)$ we obtain

$$e^{-.04t}\frac{dP}{dt} - .04e^{-.04t}P = 520e^{-.04t}$$

or

$$\frac{d(e^{-.04t}P)}{dt} = 520e^{-.04t}.$$

Integrating both sides with respect to t we obtain

$$e^{-.04t}P = \int 520e^{-.04t}\,dt$$

which gives

$$e^{-.04t}P = -13,000e^{-.04t} + c$$

so the general solution is

$$P(t) = -13,000 + ce^{.04t}.$$

Since the account starts at $500, the initial condition is $P(0) = 500$. Solving

$$500 = -13,000 + ce^{.04(0)}$$

yields $c = 13,500$. Therefore, our model is

$$P(t) = -13,000 + 13,500e^{.04t}.$$

To find the amount on deposit after 5 years, we evaluate $P(5)$ and obtain

$$-13,000 + 13,500e^{.04(5)} \approx 3,488.94.$$

25. Step 1: *Before retirement*
We first calculate how much money will be in her retirement fund after 30 years. The differential equation modeling the situation is

$$\frac{dy}{dt} = .08y + 1200,$$

where y(t) represents the fund's balance at time t. The integrating factor is $\mu(t) = e^{\int -.08\,dt} = e^{-.08t}$. Multiplying both sides of the differential equation by $\mu(t)$ yields

$$e^{-0.08t}\frac{dy}{dt} - 0.08e^{-0.08t}y = 1200e^{-.08t}.$$

Using the Product Rule on the left-hand side, we observe that this equation can be rewritten as

$$\frac{d(e^{-0.08t}y)}{dt} = 1200e^{-.08t},$$

and we integrate both sides to obtain

$$e^{-0.08t}y = \frac{1200e^{-.08t}}{-0.08} + c,$$

where c is a constant that is determined by the initial condition $y(0) = 0$. We obtain

$$y(t) = ce^{0.08t} - 15,000.$$

From the initial condition, we see that $c = 15,000$, and

$$y(t) = 15,000(e^{0.08t} - 1).$$

Letting $t = 30$, we compute that the fund contains $150,347.65 after 30 years.
Step 2: *After retirement*
We need a new model for the remaining years since the professor is withdrawing rather than depositing. Since she withdraws at a rate of $3,000 per month, we write

$$\frac{dy}{dt} = .08y - 36,000,$$

where we continue to measure time t in years. As in Step 1, the integrating factor is $\mu(t) = e^{-.08t}$. We rewrite the equation, multiply both sides by $\mu(t)$, and use the Product Rule to obtain

$$\frac{d(e^{-.08t}y)}{dt} = -36,000e^{-.08t}.$$

Integrating both sides, we have

$$e^{-.08t}y = \frac{-36,000e^{-.08t}}{-0.08} + c.$$

The general solution is

$$y(t) = 450,000(1 + ce^{.08t}).$$

In this case, we have the initial condition $y(0) = 150,347.65$ since at time $t = 0$, $y(t)$ is the amount in the fund immediately after she stops working. Solving $150,347.65 = 450,000(1 + c)$ for c yields $c = -.6659$. Therefore,

$$y(t) = 450,000(1 - .6659e^{.08t}).$$

Finally, we wish to know when her money runs out. That is, at what time t is $y(t) = 0$? Solving $450,000(1 - .6659e^{.08t}) = 0$ yields $t \approx 5.08$ years or approximately 61 months.

27. We will use the term "parts" as shorthand for the product of parts per billion of dioxin and the volume of water in the tank. Basically this product represents the total amount of dioxin in the tank. The tank initially contains 200 gallons at a concentration of 2 parts per billion, which results in 400 parts of dioxin.

Let $y(t)$ be the amount of dioxin in the tank at time t. Since water with 4 parts per billion of dioxin flows in at the rate of 5 gallons per minute, 20 parts of dioxin enter the tank each minute. Also, the volume of water in the tank at time t is $200 + 2t$, so the concentration of dioxin in the tank is $y/(200 + 2t)$. Since well-mixed water leaves the tank at the rate of 2 gallons per minute, the differential equation that represents the change in the amount of dioxin in the tank is

$$\frac{dy}{dt} = 20 - 2\left(\frac{y}{200 + 2t}\right),$$

which simplifies to

$$\frac{dy}{dt} = 20 - \left(\frac{1}{100 + t}\right)y.$$

We can rewrite this equation as

$$\frac{dy}{dt} + \left(\frac{1}{100 + t}\right)y = 20,$$

and the integrating factor is

$$\mu(t) = e^{\int (1/(100+t))\,dt} = e^{\ln(100+t)} = 100 + t.$$

Multiplying both sides by $\mu(t)$, we obtain

$$(100 + t)\frac{dy}{dt} + y = 20(100 + t),$$

which is equivalent to

$$\frac{d((100 + t)y)}{dt} = 20(100 + t)$$

by the Product Rule. Integrating both sides with respect to t, we obtain

$$(100 + t)y = 2000t + 10t^2 + c.$$

Since $y(0) = 400$, we see that $c = 40,000$. Therefore,

$$y(t) = \frac{10t^2 + 2000t + 40,000}{t + 100}.$$

The tank fills up at $t = 100$, and $y(100) = 1,700$. To express our answer in terms of concentration, we calculate $y(100)/400 = 4.25$ parts per billion.

29. (a) Let $y(t)$ be the amount of salt in the tank at time t. Since the tank is being filled at a total rate of 1 gallon per minute, the volume at time t is $V_0 + t$ and the concentration of salt in the tank is

$$\frac{y}{V_0 + t}.$$

The amount of salt entering the tank is the product of 2 gallons per minute and 0.25 pounds of salt per minute. The amount of salt leaving the tank is the product of the concentration of salt in the tank and the rate that brine is leaving. In this case, the rate is 1 gallon per minute, so the amount of salt leaving the tank is $y/(V_0 + t)$. The differential equation for $y(t)$ is

$$\frac{dy}{dt} = \frac{1}{2} - \frac{y}{V_0 + t}.$$

Since the water is initially clean, the initial condition is $y(0) = 0$.

(b) If $V_0 = 0$, the differential equation above becomes

$$\frac{dy}{dt} = \frac{1}{2} - \frac{y}{t}.$$

Note that this differential equation is undefined at $t = 0$. Thus, we *cannot* apply the Existence and Uniqueness Theorem to guarantee a unique solution at time $t = 0$. However, we can still solve the equation using our standard techniques assuming that $t \neq 0$.
Rewriting the equation as

$$\frac{dy}{dt} + \frac{y}{t} = \frac{1}{2},$$

we see that the integrating factor is

$$\mu(t) = e^{\int (1/t)\,dt} = e^{\ln t} = t.$$

Multiplying both sides of the differential equation by $\mu(t)$, we obtain

$$t\frac{dy}{dt} + y = \frac{t}{2},$$

which is equivalent to

$$\frac{d(ty)}{dt} = \frac{t}{2}.$$

Integrating both sides with respect to t, we get

$$ty = \frac{t^2}{4} + c,$$

so that the general solution is

$$y(t) = \frac{t}{4} + \frac{c}{t}.$$

Since the above expression is undefined at $t = 0$, we cannot make use of the initial condition $y(0) = 0$ to find the desired solution.
However, if the tank is initially empty, the concentration of salt in the tank remains constant over time at 0.25 pounds of salt per gallon. Therefore, we reconsider the equation

$$\frac{y}{t} = \frac{1}{4} + \frac{c}{t^2}.$$

If $c = 0$, we have $y/t = 1/4$. Hence, $c = 0$ yields the solution $y(t) = t/4$ which *is* a valid model for this situation.
It is useful to note that, if $V_0 = 0$, then we do not really need a differential equation to model the amount of the salt in the tank as a function of time. Clearly the concentration is constant as a function of time, and therefore the amount of salt in the tank is the product of the concentration and the volume of brine in the tank.

EXERCISES FOR SECTION 1.9

1. We rewrite the equation as

$$\frac{dy}{dt} = (y - 4t) + (y - 4t)^2 + 4,$$

use that

$$\frac{dy}{dt} = \frac{du}{dt} + 4,$$

and substitute to obtain

$$\frac{du}{dt} + 4 = u + u^2 + 4.$$

This equation simplifies to

$$\frac{du}{dt} = u^2 + u,$$

which is nonlinear, autonomous, and separable.

3. Rewrite the equation as

$$\frac{dy}{dt} = ty + (ty)^2 + \cos(ty).$$

Using that $y = u/t$, we have

$$\frac{dy}{dt} = \frac{1}{t}\frac{du}{dt} - \frac{1}{t^2}u.$$

We substitute to obtain

$$\frac{1}{t}\frac{du}{dt} - \frac{1}{t^2}u = u + u^2 + \cos u,$$

which simplifies to

$$\frac{du}{dt} = \frac{u}{t} + t(u + u^2 + \cos u).$$

This equation is nonlinear and nonautonomous.

   This is a good example of a change of variables that looks like it is going to greatly simplify the equation but does not because the term that replaces dy/dt is complicated.

5. Let $u = y - t$. Then

$$\frac{du}{dt} = \frac{dy}{dt} - 1,$$

and the differential equation becomes

$$\frac{du}{dt} + 1 = u^2 - u - 1,$$

which simplifies to

$$\frac{du}{dt} = u^2 - u - 2 = (u - 2)(u + 1).$$

The equilibrium points are $u = 2$ (a source) and $u = -1$ (a sink). These equilibria correspond to the solutions $y_1(t) = 2 + t$ and $y_2(t) = -1 + t$.

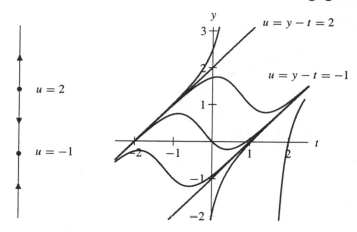

7. First we clear the denominator, so the equation becomes

$$t\frac{dy}{dt} = ty(\cos ty) - y.$$

Then we change of variables using $u = ty$. We have

$$\frac{du}{dt} = t\frac{dy}{dt} + y,$$

and the differential equation becomes

$$\frac{du}{dt} - \frac{u}{t} = u(\cos u) - \frac{u}{t}.$$

Thus we get

$$\frac{du}{dt} = u\cos u.$$

The equilibrium point $u = 0$ corresponds to the solution $y(t)$ that is constantly zero for all t, and the equilibrium points $u = \pi/2 \pm n\pi$ for $n = 0, \pm 1, \pm 2, \ldots$ correspond to solutions $y(t) = (\pi/2 \pm n\pi)/t$, for $n = 0, \pm 1, \pm 2, \ldots$. Note that the change of variables "blows up" at $t = 0$.

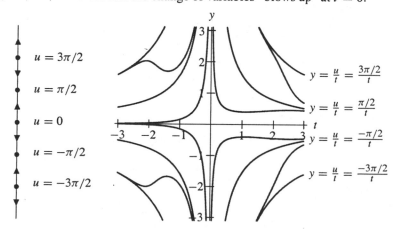

9. If $u = y/(1 + t)$, we have

$$\frac{du}{dt} = \frac{1}{1+t}\frac{dy}{dt} - \frac{y}{(1+t)^2}.$$

Then

$$\frac{dy}{dt} = (1+t)\frac{du}{dt} + u,$$

and the differential equation becomes

$$(1+t)\frac{du}{dt} + u = u - \frac{u(1+t)}{t} + t^2(t+1),$$

which reduces to

$$\frac{du}{dt} = -\frac{u}{t} + t^2.$$

This differential equation is linear, and we rewrite it as

$$\frac{du}{dt} + \frac{u}{t} = t^2.$$

Its integrating factor is

$$\mu(t) = e^{\int (1/t)\,dt} = e^{\ln t} = t.$$

Multiplying both sides of the differential equation by $\mu(t)$, we obtain

$$t\frac{du}{dt} + u = t^3,$$

which is equivalent to

$$\frac{d(tu)}{dt} = t^3.$$

Integrating both sides, we have

$$tu = \frac{t^4}{4} + c,$$

where c is an arbitrary constant. Therefore,

$$u(t) = \frac{t^3}{4} + \frac{c}{t}.$$

To determine the general solution for $y(t)$, we have $y = u(1 + t)$, and therefore

$$y(t) = \frac{t^3(1+t)}{4} + \frac{c(1+t)}{t}$$

$$= \frac{t^3(1+t)}{4} + \frac{c}{t} + c.$$

11. We know that

$$\frac{du}{dt} = \frac{dy}{dt} - 1.$$

Substituting y and dy/dt into the equation $du/dt = (1 - u)u$, we get

$$\frac{dy}{dt} - 1 = (1 - (y - t))(y - t),$$

which simplifies to

$$\frac{dy}{dt} = -y^2 + 2yt + y - t^2 - t + 1.$$

Graphs of solutions $y(t)$ are obtained from the graphs of the solutions to $du/dt = (1 - u)u$ by "tilting up" the plane so that horizontal lines in the tu-plane become lines of slope 1 in the ty-plane. As a result, the graphs of the equilibrium solutions in the tu-plane become lines with slope 1 in the ty-plane.

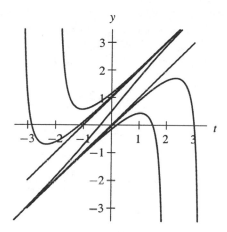

13. If $y = u^2$, then $u = \sqrt{y}$ and

$$\frac{du}{dt} = \frac{1}{2\sqrt{y}} \frac{dy}{dt}.$$

Substituting y and dy/dt into $du/dt = (1 - u)u$, we have

$$\frac{1}{2\sqrt{y}} \frac{dy}{dt} = (1 - \sqrt{y})\sqrt{y},$$

which simplifies to

$$\frac{dy}{dt} = 2y(1 - \sqrt{y}).$$

Note that the change of variables fixes the equilibrium points $u = 0$ and $u = 1$. In other words, $y = 0$ and $y = 1$ are also equilibrium points.

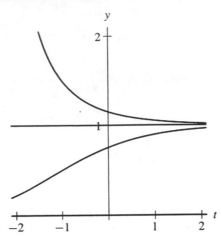

15. (a) Since 1 gallon per minute of salt water containing 2 pounds of salt per gallon and 5 gallons per minute of salt water containing 0.2 pounds of salt per gallon enter the vat, 3 pounds of salt per minute enters the vat. The concentration of the salt water at time t is $S(t)/(10 + 3t)$. Since 3 gallons per minute of salt water leaves the vat, $3S(t)/(10 + 3t)$ pounds of salt is removed. The rate of change of the amount of salt in the vat is

$$\frac{dS}{dt} = 3 - \frac{3S}{10 + 3t}.$$

(b) The concentration of the salt in the vat at time t is $C(t) = S(t)/(10 + 3t)$. Using the Product Rule for differentiation, we obtain

$$\frac{dC}{dt} = \frac{1}{10 + 3t}\frac{dS}{dt} - \frac{3C}{10 + 3t},$$

and thus

$$\frac{dS}{dt} = (10 + 3t)\frac{dC}{dt} + 3C.$$

Since $dS/dt = 3 - 3(S/(10 + 3t)) = 3 - 3C$, we obtain

$$(10 + 3t)\frac{dC}{dt} + 3C = 3 - 3C,$$

which yields

$$\frac{dC}{dt} = \frac{3 - 6C}{10 + 3t}.$$

(c) From the right-hand side of the differential equation, note that $C(t) = 1/2$ is an equilibrium solution. Moreover, for solutions whose initial conditions satisfy $C(0) < 1/2$, we see that $dC/dt > 0$. In our case, $C(0) = 0$, and we know that $C(t) \to 1/2$ as $t \to \infty$.

(d) From the differential equation in part (b), we separate variables and obtain

$$\frac{dC}{3 - 6C} = \frac{dt}{10 + 3t}.$$

Integration yields

$$C = \frac{1}{2} - \frac{k}{(10 + 3t)^2}.$$

At $t = 0$, $C(t) = 0$. Therefore, we have $0 = 1/2 - k/100$. Hence $k = 50$, and

$$C(t) = \frac{1}{2} - \frac{50}{(10 + 3t)^2}.$$

Evaluating this expression at $t = 5$ yields $C(5) = 0.42$.

17. **(a)** To find the equilibrium points, we solve $dy/dt = 0$, which is equivalent to

$$10y^3 = 1.$$

We get $y = \sqrt[-3]{10}$.

(b) Let $u = y - \sqrt[-3]{10}$. Then, $du/dt = dy/dt$, and

$$10y^3 - 1 = 10\left(u + \sqrt[-3]{10}\right)^3 - 1$$

$$= 10\left(u^3 + 3\sqrt[-3]{10}\,u^2 + 3\sqrt[-3]{10^2}\,u + \frac{1}{10}\right) - 1$$

$$= 10u^3 + 3 \cdot 10^{2/3}u^2 + 3\sqrt[3]{10}\,u.$$

The linear approximation is

$$\frac{du}{dt} = 3\sqrt[3]{10}\,u.$$

Since $3\sqrt[3]{10} > 0$, the origin is a source.

(c) Solving the linear system, we get

$$u(t) = u_0 e^{3\sqrt[3]{10}\,t}$$

near the origin. In other words, solutions move away from the equilibrium solution exponentially fast with exponent $3\sqrt[3]{10}\,t$. In particular, to derive the amount of time that it takes a solution to double its distance from the equilibrium, we solve

$$2u_0 = u_0 e^{3\sqrt[3]{10}\,t}$$

and obtain $t = (\ln 2)/(3\sqrt[3]{10})$.

19. **(a)** To find the equilibria, we solve $dy/dt = 0$ and obtain $y = -1$ and $y = 3$.

(b) For $y = -1$, let $u = y + 1$. Then, $du/dt = dy/dt$ and

$$\frac{du}{dt} = \frac{dy}{dt} = (y + 1)(3 - y) = u(4 - u).$$

The linear approximation is

$$\frac{du}{dt} = 4u,$$

so $u = 0$ is a source.

Similarly, for $y = 3$, we let $v = y - 3$ and obtain

$$\frac{dv}{dt} = -v(v + 4) = -4v - v^2.$$

The linear approximation is

$$\frac{dv}{dt} = -4v.$$

so $v = 0$ is a sink.

(c) The linearizations at $y = -1$ and $y = 3$ have general solutions $u = u_0 e^{4t}$ and $v = v_0 e^{-4t}$, respectively. Hence, the time necessary to halve or double the distance to the equilibrium point starting nearby is $t = (\ln 2)/4$.

21. (a) The linearization of $\sin y$ about $y = 0$ is y. Therefore, the linearization of this differential equation near the equilibrium point at $y = 0$ is

$$\frac{dy}{dt} = -3y.$$

(b) The general solution of the linearized differential equation is

$$y = y_0 e^{-3t},$$

where y_0 is the initial condition $y(0)$. The solution to the linearized equation with $y(0) = 0.04$ is $0.04e^{-3t}$, and therefore, $y(2) \approx 0.04e^{-6} \approx 0.000099$.

23. (a) If $u = P - 100$, the differential equation becomes

$$\frac{du}{dt} = 0.05(u + 100) \left(\frac{-u}{100} \right) - 0.02(1 + \sin t)$$

$$= -0.05u \left(\frac{u}{100} + 1 \right) - 0.02(1 + \sin t).$$

(b) If we linearize the autonomous part of the differential equation near $u = 0$, we get

$$\frac{du}{dt} = -0.05u - 0.02(1 + \sin t).$$

(c) This linearized equation is nonhomogeneous, and we can find the solutions using an integrating factor (for an alternate method, see Appendix A). We rewrite the equation as

$$\frac{du}{dt} + 0.05u = -0.02(1 + \sin t),$$

and the integrating factor is $\mu(t) = e^{0.05t}$. Multiplying both sides by $\mu(t)$, we get

$$\frac{d(\mu \cdot u)}{dt} = e^{0.05t}(-0.02(1 + \sin t)).$$

We integrate both sides and get

$$\mu \cdot u = \int -0.02(1 + \sin t)e^{0.05t}\, dt,$$

which yields the general solution

$$u(t) = -0.4 - \tfrac{2}{2005} \sin t + \tfrac{8}{401} \cos t + ce^{-0.05t}.$$

(We did the integral exactly with a little help from our computer algebra system.)

(d) For large t, $ce^{-0.05t} \to 0$, and $u(t)$ fluctuates about -0.4, and the amplitude of these fluctuations is slightly less than 0.02 in both directions. Thus, $u(t)$ oscillates between -0.38 and -0.42, and we estimate that P fluctuates about 99.6 with oscillations that are roughly 0.02 in amplitude. (In Section 4.4, we show how to use complex arithmetic to estimate the amplitude of these oscillations.)

25. (a) Since $y = u + y_0$, $du/dt = dy/dt$, and we can replace y by $u + y_0$ in the right-hand side of the differential equation. We get

$$\frac{du}{dt} = f(u + y_0).$$

(b) Using the Taylor series

$$f(y) = f(y_0) + f'(y_0)(y - y_0) + \frac{f''(y_0)}{2!}(y - y_0)^2 + \dots$$

about $y = y_0$, we have

$$\frac{du}{dt} = f(y_0) + f'(y_0)\,u + \frac{f''(y_0)}{2!}\,u^2 + \dots$$

because $u = y - y_0$. Since y_0 is an equilibrium point, $f(y_0) = 0$. Thus, when we truncate higher-order terms (order ≥ 2) in u, we get the linearized equation

$$\frac{du}{dt} = f'(y_0)\,u.$$

CHAPTER 2

First-Order Systems

EXERCISES FOR SECTION 2.1

1. In the case where it takes many predators to eat one prey, the constant in the negative effect term of predators on the prey is small. Therefore, (ii) corresponds the system of large prey and small predators. On the other hand, one predator eats many prey for the system of large predators and small prey, and, therefore, the coefficient of negative effect term on predator-prey interaction on the prey is large. Hence, (i) corresponds to the system of small prey and large predators.

3. Substitution of $y = 0$ into the equation for dy/dt yields $dy/dt = 0$ for all t. Therefore, $y(t)$ is constant, and since $y(0) = 0$, $y(t) = 0$ for all t.

   Note that to verify this assertion rigorously, we need a uniqueness theorem (see Section 2.4).

5. Substitution of $x = 0$ into the equation for dx/dt yields $dx/dt = 0$ for all t. Therefore, $x(t)$ is constant, and since $x(0) = 0$, $x(t) = 0$ for all t.

   Note that to verify this assertion rigorously, we need a uniqueness theorem (see Section 2.4).

7. The population starts with a relatively large rabbit (R) and a relatively small fox (F) population. The rabbit population grows, then the fox population grows while the rabbit population decreases. Next the fox population decreases until both populations are close to zero. Then the rabbit population grows again and the cycle starts over. Each repeat of the cycle is less dramatic (smaller total oscillation) and both populations oscillate toward an equilibrium which is approximately $(R, F) = (1/2, 3/2)$.

9. By hunting, the number of prey decreases α units per unit of time. Therefore, the rate of change dR/dt of the number of prey has the term $-\alpha$. Only the equation for dR/dt needs modification.

   (i) $dR/dt = 2R - 1.2RF - \alpha$

   (ii) $dR/dt = R(2 - R) - 1.2RF - \alpha$

11. Since the second food source is unlimited, if $R = 0$ and k is the growth parameter for the predator population, F obeys an exponential growth model, $dF/dt = kF$. The only change we have to make is in the rate of F, dF/dt. For both (i) and (ii), $dF/dt = kF + 0.9RF$.

13. If $R - 5F > 0$, the number of predators increases and, if $R - 5F < 0$, the number of predators decreases. Since the condition on prey is same, we modify only the predator part of the system. the modified rate of change of the predator population is

$$\frac{dF}{dt} = -F + 0.9RF + k(R - 5F)$$

where $k > 0$ is the immigration parameter for the predator population.

15. Suppose $y = 1$. If we can find a value of x such that $dy/dt = 0$, then for this x and $y = 1$ the predator population is constant. (This point may not be an equilibrium point because we do not know if $dx/dt = 0$.) The required value of x is $x = 0.05$ in system (i) and $x = 20$ in system (ii). Survival for one unit of predators requires 0.05 units of prey in (i) and 20 units of prey in (ii). Therefore, (i) is a system of inefficient predators and (ii) is a system of efficient predators.

17. **(a)** For the initial condition close to zero, the pest population increases much more rapidly than the predator. After a sufficient increase in the predator population, the pest population starts to

decrease while the predator population keeps increasing. After a sufficient decrease in the pest population, the predator population starts to decrease. Then, the population comes back to the initial point.

(b) After applying the pest control, you may see the increase of the pest population due to the absence of the predator. So in the short run, this sort of pesticide can cause an explosion in the pest population.

19. (a) Substituting $y(t) = \sin t$ into the left-hand side of the differential equation gives

$$\frac{d^2 y}{dt^2} + y = \frac{d^2 (\sin t)}{dt^2} + \sin t$$

$$= -\sin t + \sin t$$

$$= 0,$$

(b)

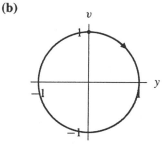

so the left-hand side equals the right-hand side for all t.

(c) These two solutions trace the same curve in the yv-plane—the unit circle.

(d) The difference in the two solution curves is in how they are parameterized. The solution in this problem is at $(0, 1)$ at time $t = 0$ and hence it lags behind the solution in the section by $\pi/2$. This information cannot be observed solely by looking at the solution curve in the phase plane.

21. Hooke's law tells us that the restoring force exerted by a spring is linearly proportional to the spring's displacement from its rest position. In this case, the displacement is 3 in. while the restoring force is 12 lbs. Therefore, 12 lbs. $= k \cdot 3$ in. or $k = 4$ lbs. per in. $= 48$ lbs. per ft.

23. An extra firm mattress does not deform when you lay on it. This means that it takes a great deal of force to compress the springs so the spring constant must be large.

25. Suppose $\alpha > 0$ is the reaction rate constant for A+B $\rightarrow$ C. The reaction rate is αab at time t, and after the reaction, a and b decrease by αab. We therefore obtain the system

$$\frac{da}{dt} = -\alpha ab$$

$$\frac{db}{dt} = -\alpha ab.$$

27. Suppose k_1 and k_2 are the rates of increase of A and B respectively. Since A and B are added to the solution at constant rates, k_1 and k_2 are added to da/dt and db/dt respectively. The system becomes

$$\frac{da}{dt} = k_1 - \alpha ab$$

$$\frac{db}{dt} = k_2 - \alpha ab.$$

29. Suppose γ is the reaction-rate coefficient for the reaction $B + B \rightarrow A$. By the reaction, two B's react with each other to create one A. In other words, B decreases at the rate γb^2 and A increases at the rate $\gamma b^2/2$. The resulting system of the differential equations is

$$\frac{da}{dt} = k_1 - \alpha ab + \frac{\gamma b^2}{2}$$

$$\frac{db}{dt} = k_2 - \alpha ab - \gamma b^2.$$

EXERCISES FOR SECTION 2.2

1. (a) $\mathbf{V}(x, y) = (1, 0)$ **(b)** See part (c).

    (c)   **(d)**

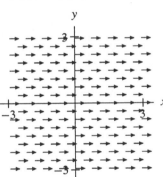
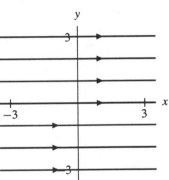

    (e) As t increases, solutions move along horizontal lines toward the right.

3. (a) $\mathbf{V}(y, v) = (-v, y)$ **(b)** See part (c).

    (c)   **(d)**

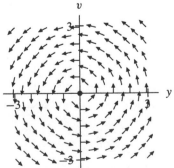
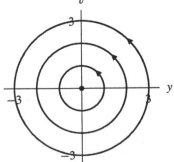

    (e) As t increases, solutions move on circles around $(0, 0)$ in the counter-clockwise direction.

5. (a) $\mathbf{V}(x, y) = (x, -y)$ **(b)** See part (c).

(c)

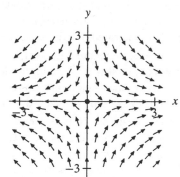

(d)

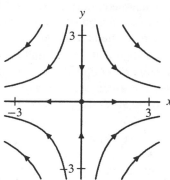

(e) As t increases, solutions move toward the x-axis in the y-direction and away from the y-axis in the x-direction.

7. (a) Let $v = dy/dt$. Then

(b) See part (c).

$$\frac{dv}{dt} = \frac{d^2y}{dt^2} = y.$$

Thus the associated vector field is $\mathbf{V}(y, v) = (v, y)$.

(c)

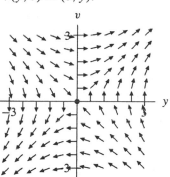

(d)

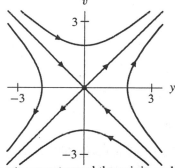

(e) As t increases, solutions in the 2nd and 4th quadrants move toward the origin and away from the line $y = -v$. Solutions in the 1st and 3rd quadrants move away from the origin and toward the line $y = v$.

9. (a)

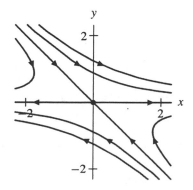

(b) The solution tends to the origin along the line $y = -x$ in the xy-phase plane. Therefore both $x(t)$ and $y(t)$ tend to zero as $t \to \infty$.

11. **(a)** There are equilibrium points at $(\pm 1, 0)$, so only systems (ii) and (vii) are possible. Since the direction field points away from the origin along the y-axis, the equation $dy/dt = -y$ does not match this field. Therefore, system (ii) is the system that generated this direction field.

   (b) The origin is the only equilibrium point, so the possible systems are (iii), (iv), (v), and (viii). Vectors point directly away from the origin on the y-axis, so this direction field does not correspond to systems (iii) and (viii). Along the line $y = x$, the vectors are more vertical than horizontal. Consequently this direction field corresponds to system (v) rather than system (iv).

   (c) The origin is the only equilibrium point, so the possible systems are (iii), (iv), (v), and (viii). The direction field is not tangent to the y-axis, so it does not match either system (iv) or (v). Vectors point toward the origin on the line $y = x$, so $dy/dt = dx/dt$ if $y = x$. This condition is not satisfied by system (iii). Therefore, this direction field corresponds to system (viii).

   (d) The only equilibrium point is $(0, 1)$. System (i) is the only system with a unique equilibrium point at $(0, 1)$.

13. The $x(t)$- and $y(t)$-graphs are periodic, so they correspond to a solution curve that returns to its initial condition in the phase plane. In other words, its solution curve is a closed curve. Since the amplitude of the oscillation of $x(t)$ is relatively large, these graphs must correspond to the outermost closed solution curve.

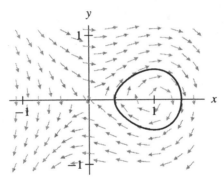

15. The graphs are not periodic, so they cannot correspond to the two closed solution curves in the phase portrait. Only one graph crosses the t-axis. The other graph remains negative for all time. Note that the two graphs cross.

   The corresponding solution curve is the one that starts in the second quadrant and crosses the x-axis and the line $y = x$ as it moves through the third quadrant.

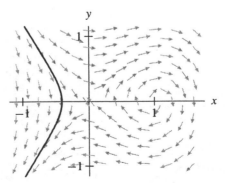

17. **(a)** To find the equilibrium points, we solve the system of equations

$$\begin{cases} 4x - 7y + 2 = 0 \\ 3x + 6y - 1 = 0. \end{cases}$$

These simultaneous equations have one solution, $(x, y) = (-1/9, 2/9)$.

(b)

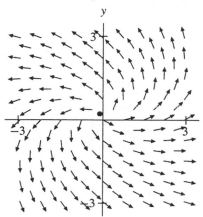

(c) As t increases, typical solutions spiral away from the origin in the counter-clockwise direction.

19. **(a)** To find the equilibrium points, we solve the system of equations

$$\begin{cases} \cos w = 0 \\ -z + w = 0. \end{cases}$$

The first equation implies that $w = \pi/2 + k\pi$ where k is any integer, and the second equation implies that $z = w$. The equilibrium points are $(\pi/2 + k\pi, \pi/2 + k\pi)$ for any integer k.

(b)

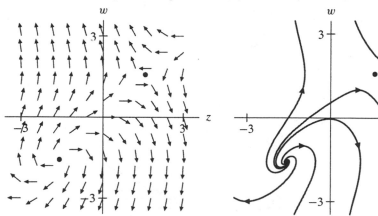

(c) As t increases, typical solutions move away from the line $z = w$, which contains the equilibrium points. The value of w is either increasing or decreasing without bound depending on the initial condition.

21. **(a)** To find the equilibrium points, we solve the system of equations

$$\begin{cases} y = 0 \\ -\cos x - y = 0. \end{cases}$$

We see that $y = 0$, and thus $\cos x = 0$. The equilibrium points are $(\pi/2 + k\pi, 0)$ for any integer k.

(b)

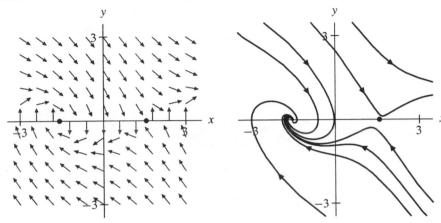

(c) As t increases, typical solutions spiral toward one of the equilibria on the x-axis. Which equilibrium point the solution approaches depends on the initial condition.

23. Since the solution curve spirals into the origin, the corresponding $x(t)$- and $y(t)$-graphs must oscillate about the t-axis with the decreasing amplitudes.

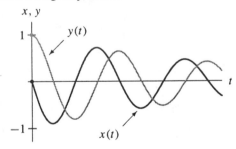

25. The $x(t)$-graph satisfies $-2 < x(0) < -1$ and increases as t increases. The $y(t)$-graph satisfies $1 < y(0) < 2$. Initially it decreases until it reaches its minimum value of $y = 1$ when $x = 0$. Then it increases as t increases.

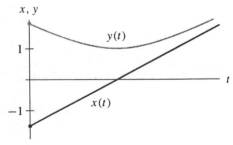

27. From the graphs, we see that $y(0) = 0$ and $x(0)$ is slightly positive. Initially both graphs increase. Then they cross, and slightly later $x(t)$ attains its maximum value. Continuing along we see that $y(t)$ attains its maximum at the same time as $x(t)$ crosses the t-axis.

In the xy-phase plane these graphs correspond to a solution curve that starts on the positive x-axis, enters the first quadrant, crosses the line $y = x$, and eventually crosses the y-axis into the second quadrant exactly when $y(t)$ assumes its maximum value. For this portion of the curve, $y(t)$ is increasing while $x(t)$ assumes a maximum and starts decreasing.

We see that once $y(t)$ attains its maximum, it decreases for a prolonged period of time until it assumes its minimum value. Throughout this interval, $x(t)$ remains negative although it assumes its minimum value twice and a local maximum value once. In the phase plane, the solution curve enters the second quadrant and then crosses into the third quadrant when $y(t) = 0$. The $x(t)$- and $y(t)$-graphs cross precisely when the solution curve crosses the line $y = x$ in the third quadrant.

Finally the $y(t)$-graph is increasing again while the $x(t)$-graph becomes positive and assumes its maximum value once more. The two graphs return to their initial values. In the phase plane this behavior corresponds to the solution curve moving from the third quadrant through the fourth quadrant and back to the original starting point.

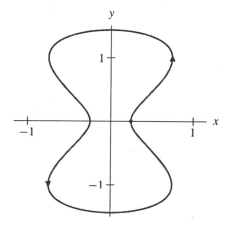

29. Since the vector field does not change with time, Gib will follow the same path as Harry, only one time unit behind. See Section 2.4 for the Uniqueness Theorem that rigorously justifies these statements.

31. Often the solutions in the quiz are over a longer time interval than what is shown in the following graphs.

(a)   **(b)**

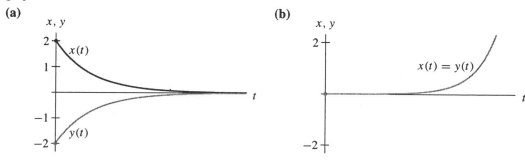

(c)

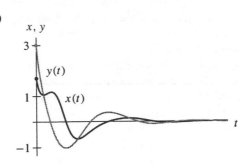

(d)

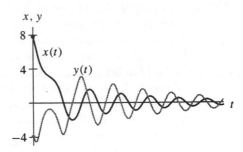

(e)

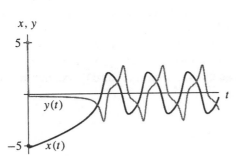

(f)

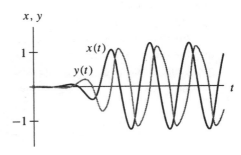

(g)

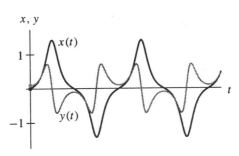

(h)

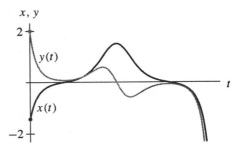

(i)

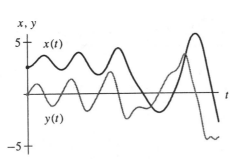

EXERCISES FOR SECTION 2.3

1. To check that $dx/dt = 2x + 2y$, we compute both

$$\frac{dx}{dt} = 2e^t$$

and

$$2x + 2y = 4e^t - 2e^t = 2e^t.$$

To check that $dy/dt = x + 3y$, we compute both

$$\frac{dy}{dt} = -e^t,$$

and

$$x + 3y = 2e^t - 3e^t = -e^t.$$

Both equations are satisfied for all t. Hence $(x(t), y(t))$ is a solution.

3. To check that $dx/dt = 2x + 2y$, we compute both

$$\frac{dx}{dt} = 2e^t - 4e^{4t}$$

and

$$2x + 2y = 4e^t - 2e^{4t} - 2e^t + 2e^{4t} = 2e^t.$$

Since the results of these two calculations do not agree, the first equation in the system is not satisfied, and $(x(t), y(t))$ is not a solution.

5. The second equation in the system is $dy/dt = -y$, and from Section 1.1, we know that $y(t)$ must be a function of the form $y_0 e^{-t}$, where y_0 is the initial value.

7. From the second equation, we know that $y(t) = k_1 e^{-t}$ for some constant k_1. Using this observation, the first equation in the system can be rewritten as

$$\frac{dx}{dt} = 2x + k_1 e^{-t}.$$

This equation is a first-order linear equation, and we can derive the general solution using integrating factors from Section 1.8 or using the Extended Linearity Principle from Appendix A.

For this equation the integrating factor is $\mu = e^{-2t}$. If we begin with the equation

$$\frac{dx}{dt} - 2x = k_1 e^{-t}$$

and multiply both sides by μ, we obtain

$$e^{-2t} \left(\frac{dx}{dt} - 2x \right) = e^{-2t} \left(k_1 e^{-t} \right),$$

which reduces to

$$\frac{d}{dt} \left(e^{-2t} x \right) = k_1 e^{-3t}.$$

Integrating both sides, we have

$$e^{-2t}x = -\frac{k_1}{3}e^{-3t} + k_2,$$

where k_2 is a constant of integration. Multiplying both sides by e^{2t}, we obtain

$$x(t) = -\frac{k_1}{3}e^{-t} + k_2 e^{2t}.$$

9. (a) Given the general solution

$$\left(k_2 e^{2t} - \frac{k_1}{3}e^{-t}, k_1 e^{-t} \right),$$

we see that $k_1 = 0$, and therefore $k_2 = 1$. We obtain $\mathbf{Y}(t) = (x(t), y(t)) = (e^{2t}, 0)$.

(b)

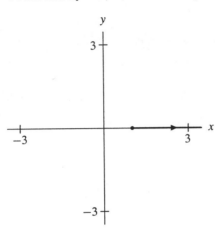

(c)

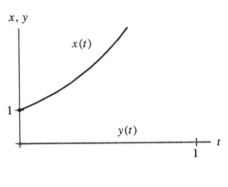

11. (a) Given the general solution

$$\left(k_2 e^{2t} - \frac{k_1}{3}e^{-t}, k_1 e^{-t} \right),$$

we see that $k_1 = 1$, and therefore $k_2 = 1/3$. We obtain

$$\mathbf{Y}(t) = (x(t), y(t)) = \left(\frac{1}{3}e^{2t} - \frac{1}{3}e^{-t}, e^{-t} \right).$$

(b)

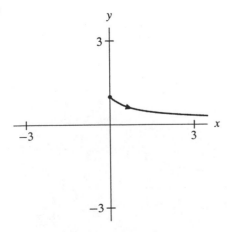

(c)

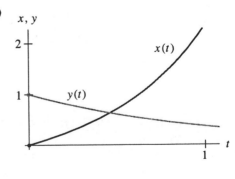

13. We choose the left wall to be the position $y = 0$ with $y > 0$ indicating positions to the right. Each spring exerts a force on the mass. If the position of the mass is y, then the left spring is stretched by the amount $y - L_1$. Therefore, the force F_1 exerted by this spring is

$$F_1 = k_1 (L_1 - y).$$

Similarly, the right spring is stretched by the amount $(1 - y) - L_2$. However, the restoring force F_2 of the right spring acts in the direction of increasing values of y. Therefore, we have

$$F_2 = k_2 ((1 - y) - L_2).$$

Using Newton's second law, we have

$$m\frac{d^2y}{dt^2} = k_1 (L_1 - y) + k_2 ((1 - y) - L_2) - b\frac{dy}{dt},$$

where the term involving dy/dt represents the force due to damping. After a little algebra, we obtain

$$m\frac{d^2y}{dt^2} + b\frac{dy}{dt} + (k_1 + k_2)y = k_1L_1 - k_2L_2 + k_2.$$

15. (a) See part (c).

(b) We guess that there are solutions of the form $y(t) = e^{st}$ for some choice of the constant s. To determine these values of s, we substitute $y(t) = e^{st}$ into the left-hand side of the differential equation, obtaining

$$\frac{d^2y}{dt^2} + 3\frac{dy}{dt} - 10y = \frac{d^2(e^{st})}{dt^2} + 3\frac{d(e^{st})}{dt} - 10(e^{st})$$

$$= s^2e^{st} + 3se^{st} - 10e^{st}$$

$$= (s^2 + 3s - 10)e^{st}$$

In order for $y(t) = e^{st}$ to be a solution, this expression must be 0 for all t. In other words,

$$s^2 + 3s - 10 = 0.$$

This equation is satisfied only if $s = -5$ or $s = 2$. We obtain two solutions, $y_1(t) = e^{-5t}$ and $y_2(t) = e^{2t}$, of this equation.

(c)

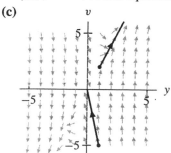
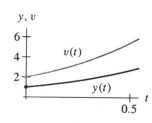
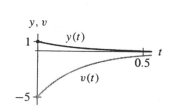

17. (a) See part (c).

(b) We guess that there are solutions of the form $y(t) = e^{st}$ for some choice of the constant s. To determine these values of s, we substitute $y(t) = e^{st}$ into the left-hand side of the differential equation, obtaining

$$\frac{d^2y}{dt^2} + 3\frac{dy}{dt} + 2y = \frac{d^2(e^{st})}{dt^2} + 4\frac{d(e^{st})}{dt} + e^{st}$$

$$= s^2 e^{st} + 4s e^{st} + e^{st}$$

$$= (s^2 + 4s + 1)e^{st}$$

In order for $y(t) = e^{st}$ to be a solution, this expression must be 0 for all t. In other words,

$$s^2 + 4s + 1 = 0.$$

Applying the quadratic formula, we obtain the roots $s = -2 \pm \sqrt{3}$ and the two solutions, $y_1(t) = e^{(-2-\sqrt{3})t}$ and $y_2(t) = e^{(-2+\sqrt{3})t}$, of this equation.

(c)

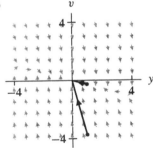
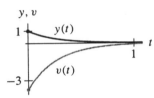

19. (a) For this system, we note that the equation for dy/dt depends only on y. In fact this equation is separable and linear, so we have a choice of techniques for finding the general solution. The general solution for y is

$$y(t) = -1 + k_1 e^t,$$

where k_1 can be any constant.

Substituting $y = -1 + k_1 e^t$ into the equation for dx/dt, we have

$$\frac{dx}{dt} = (-1 + k_1 e^t)x.$$

This equation is also separable and linear, and we find by integration that the general solution is

$$x(t) = k_2 e^{-t + k_1 e^t},$$

where k_2 is any constant. The general solution for the system is therefore

$$(x(t), y(t)) = (k_2 e^{-t + k_1 e^t}, -1 + k_1 e^t),$$

where k_1 and k_2 are constants which we can adjust to satisfy any given initial condition.

(b) Setting $dy/dt = 0$, we obtain $y = -1$. From $dx/dt = xy = 0$, we see that $x = 0$. Therefore, this system has exactly one equilibrium point, $(x, y) = (0, -1)$.

(c) If $(x(0), y(0)) = (1, 0)$, then we must solve the simultaneous equations

$$\begin{cases} k_2 e^{-t+k_1 e^t} = 1 \\ -1 + k_1 e^t = 0. \end{cases}$$

Hence $k_1 = 1$ and $k_2 e^1 = 1$. In other words, $k_2 = 1/e = e^{-1}$. Thus the solution satisfying this initial condition is

$$(x(t), y(t)) = \left(e^{-1}e^{-t+e^t}, -1 + e^t\right) = (e^{-1-t+e^t}, -1 + e^t).$$

(d)

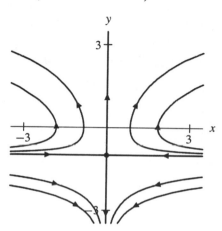

EXERCISES FOR SECTION 2.4

1. **(a)** We compute

$$\frac{dx}{dt} = \frac{d(\cos t)}{dt} = -\sin t = -y \quad \text{and} \quad \frac{dy}{dt} = \frac{d(\sin t)}{dt} = \cos t = x,$$

so $(\cos t, \sin t)$ is a solution.

(b)

Table 2.1

t	Euler's approx.	actual	distance
0	$(1, 0)$	$(1, 0)$	
4	$(-2.06, -1.31)$	$(-0.65, -0.76)$	1.51
6	$(2.87, -2.51)$	$(0.96, -0.28)$	2.94
10	$(-9.21, 1.41)$	$(-0.84, -0.54)$	8.59

(c)

Table 2.2

t	Euler's approx.	actual	distance
0	$(1, 0)$	$(1, 0)$	
4	$(-.81, -.91)$	$(-0.65, -0.76)$	0.22
6	$(1.29, -.40)$	$(0.96, -0.28)$	0.35
10	$(-1.41, -.85)$	$(-0.84, -.54)$	0.65

(d) The solution curves for this system are all circles centered at the origin. Since Euler's method uses tangent lines to approximate the solution curve and the tangent line to any point on a circle is entirely outside the circle (except at the point of tangency), each step of the Euler approximation takes the approximate solution farther from the origin. So the Euler approximations always spiral away from the origin for this system.

3. **(a)** Euler approximation yields $(x_5, y_5) \approx (0.65, -0.59)$.

(b)

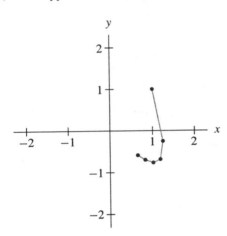

(c)

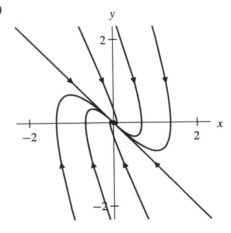

5. **(a)** Euler approximation yields $(x_5, y_5) \approx (1.94, -0.72)$.

(b)

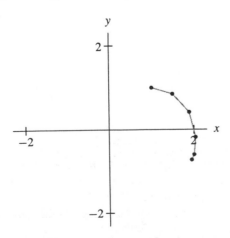

(c)

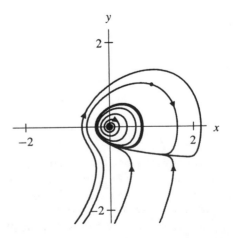

7. With $x(t) = e^{-t} \sin(3t)$ and $y(t) = e^{-t} \cos(3t)$, we have

$$\frac{dx}{dt} = -e^{-t} \sin(3t) + 3e^{-t} \cos(3t)$$

$$= -x + 3y$$

$$\frac{dy}{dt} = -3e^{-t} \sin(3t) - e^{-t} \cos(3t)$$

$$= -3x - y$$

Therefore, $\mathbf{Y}_1(t)$ is a solution.

9. If we express the solution curve for $\mathbf{Y}_1(t)$ in terms of polar coordinates (r, θ) where $r^2 = x^2 + y^2$, we obtain $r = e^{-t}$, and

$$\frac{x(t)}{y(t)} = \frac{e^{-t} \sin 3t}{e^{-t} \cos 3t} = \tan 3t.$$

At the same time,

$$\frac{x(t)}{y(t)} = \tan \phi,$$

where $\phi = \pi/2 - \theta$. Therefore, $\tan 3t = \tan \phi$, and $3t = \pi/2 - \theta$. In other words, the angle θ changes according to the relationship $\theta = \pi/2 - 3t$.

These two computations imply that the solution curve for $\mathbf{Y}_1(t)$ spirals into the origin in a clockwise direction.

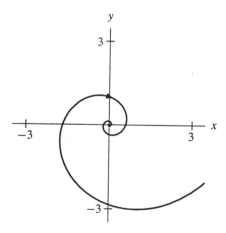

For $\mathbf{Y}_2(t)$, these computations lead to the same conclusions except that, wherever a t appears, it is replaced by $t - 1$. Thus the solution curve for $\mathbf{Y}_2(t)$ is identical to the solution curve for $\mathbf{Y}_1(t)$. In other words, as curves in the plane, they sweep out the same sets of points, but $\mathbf{Y}_1(t - 1) = \mathbf{Y}_2(t)$. Whenever $\mathbf{Y}_2(t)$ occupies a certain point in the plane, it turns out that $\mathbf{Y}_1(t)$ occupied that point exactly 1 unit of time earlier. Since these curves never occupy the same point at the same time, they do not violate the Uniqueness Theorem.

11. In order to be able to apply Euler's method to this second-order equation, we reduce the equation to a first-order system using $v = dy/dt$. We obtain

$$\frac{dy}{dt} = v$$

$$\frac{dv}{dt} = -y - \frac{v}{5}.$$

The choice of Δt has an important effect on the long-term behavior of the approximate solution curve. The curve for $\Delta t = 0.25$ spirals away from the origin. If $(y_0, v_0) = (0, 1)$, then we obtain $(y_5, v_5) \approx (0.98, 0.23)$, $(y_{10}, v_{10}) \approx (0.64, -0.92)$, $(y_{15}, v_{15}) \approx (-0.63, -0.84)$, ...

The behavior of this approximate solution curve is deceiving. Consider the approximation we obtain if we halve that value of Δt. In other words, let $\Delta t = 0.125$. For $(y_0, v_0) = (2, 0)$, then we obtain $(y_5, v_5) \approx (0.58, 0.73)$, $(y_{10}, v_{10}) \approx (0.91, 0.21)$, $(y_{15}, v_{15}) \approx (0.89, -0.37)$, ...

The following figure illustrates how this approximate solution curve spirals toward the origin. (As we will see, this second approximation is much better than the first.)

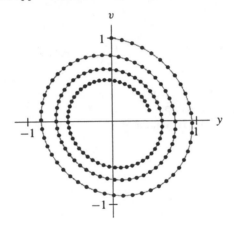

13. Assume the vector field satisfies the hypotheses of the Uniqueness Theorem. Since the vector field does not change with time, Gib will follow the same path as Harry, only one time unit behind.

15. From Exercise 14 we know that $Y_1(t-1)$ is a solution of the system and $Y_1(1-1) = Y_1(0) = Y_2(1)$, so both $Y_2(t)$ and $Y_1(t - 1)$ occupy the point $Y_1(0)$ at time $t = 1$. Hence, by the Uniqueness Theorem, they are the same solution. So $Y_2(t)$ is a reparameterization by a constant time shift of $Y_1(t)$.

EXERCISES FOR SECTION 2.5

1. **(a)** Substitution of $(0, 0, 0)$ into the given system of differential equations yields $dx/dt = dy/dt = dz/dt = 0$. Similarly, for the case of $(\pm 6\sqrt{2}, \pm 6\sqrt{2}, 27)$, we obtain

$$\frac{dx}{dt} = 10(\pm 6\sqrt{2} - (\pm 6\sqrt{2}))$$

$$\frac{dy}{dt} = 28(\pm 6\sqrt{2}) - (\pm 6\sqrt{2}) - 27(\pm 6\sqrt{2})$$

$$\frac{dz}{dt} = -\frac{8}{3}(27) - (\pm 6\sqrt{2})^2.$$

Therefore, $dx/dt = dy/dt = dz/dt = 0$, and these three points are equilibrium points.

(b) For equilibrium points, we must have $dx/dt = dy/dt = dz/dt = 0$. We therefore obtain the three simultaneous equations

$$\begin{cases} 10(y - x) = 0 \\ 28x - y - xz = 0 \\ -\frac{8}{3}z + xy = 0. \end{cases}$$

From the first equation, $x = y$. Eliminating y, we obtain

$$\begin{cases} x(27 - z) = 0 \\ -\frac{8}{3}z + x^2 = 0 \end{cases}$$

Then, $x = 0$ or $z = 27$. With $x = 0$, $z = 0$. With $z = 27$, $x^2 = 72$, hence $y = x = \pm 6\sqrt{2}$.

3. (a) We have

$$\frac{dx}{dt} = 10(y - x) = 0 \quad \text{and} \quad \frac{dy}{dt} = 28x - y = 0,$$

so $x(t) = y(t) = 0$ for all t if $x(0) = y(0) = 0$.

(b) We have

$$\frac{dz}{dt} = -\frac{8}{3}z,$$

so $z(t) = ce^{-8t/3}$. Since $z(0) = 1$, it follows that $c = 1$, and the solution is $x(t) = 0$, $y(t) = 0$, and $z(t) = e^{-8t/3}$.

(c) If $z(0) = z_0$, it follows that $c = z_0$, so the solution is $x(t) = 0$, $y(t) = 0$, and $z(t) = z_0 e^{-8t/3}$.

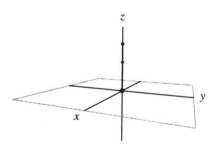

5. (a)

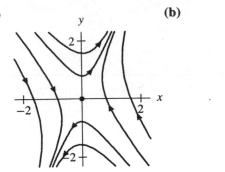

(b)

(c)

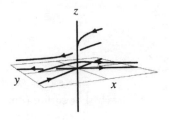

CHAPTER 3

Linear Systems

EXERCISES FOR SECTION 3.1

1. Since $a > 0$, Paul's making a profit ($x > 0$) has a beneficial effect on Paul's profits in the future because the ax term makes a positive contribution to dx/dt. However, since $b < 0$, Bob's making a profit ($y > 0$) hinders Paul's ability to make profit because the by term contributes negatively to dx/dt. Roughly speaking, business is good for Paul if his store is profitable and Bob's is not. In fact, since $dx/dt = x - y$, Paul's profits will increase whenever his store is more profitable than Bob's.

Even though $dx/dt = dy/dt = x - y$ for this choice of parameters, the interpretation of the equation is exactly the opposite from Bob's point of view. Since $d < 0$, Bob's future profits are hurt whenever he is profitable because $dy < 0$. But Bob's profits are helped whenever Paul is profitable since $cx > 0$. Once again, since $dy/dt = x - y$, Bob's profits will increase whenever Paul's store is more profitable than his.

Finally, note that both x and y change by identical amounts since dx/dt and dy/dt are always equal.

3. Since $a = 1$ and $b = 0$, we have $dx/dt = x$. Hence, if Paul is making a profit ($x > 0$), then those profits will increase since dx/dt is positive. However, Bob's profits have no effect on Paul's profits. (Note that $dx/dt = x$ is the standard exponential growth model.)

Since $c = 2$ and $d = 1$, profits from both stores have a positive effect on Bob's profits. In some sense, Paul's profits have twice the impact of Bob's profits on dy/dt.

5. $Y = \begin{pmatrix} x \\ y \end{pmatrix}$, $\dfrac{dY}{dt} = \begin{pmatrix} 2 & 1 \\ 1 & 1 \end{pmatrix} Y$

7. $Y = \begin{pmatrix} p \\ q \\ r \end{pmatrix}$, $\dfrac{dY}{dt} = \begin{pmatrix} 3 & -2 & -7 \\ -2 & 0 & 6 \\ 0 & 7.3 & 2 \end{pmatrix} Y$

9. $\dfrac{dx}{dt} = \beta y$

$\dfrac{dy}{dt} = \gamma x - y$

11. (a)   **(b)**   **(c)**

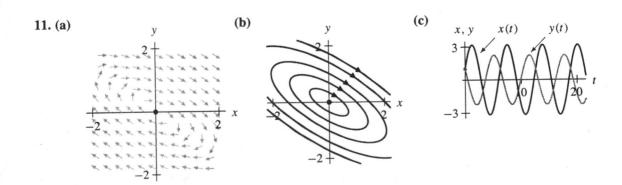

13. (a)

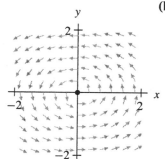

(b)

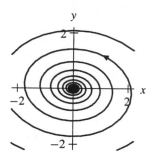

(c)

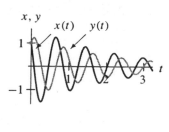

15. The vector field at a point (x_0, y_0) is (ax_0+by_0, cx_0+dy_0), so in order for a point to be an equilibrium point, it must be a solution to the system of simultaneous linear equations

$$\begin{cases} ax_0 + by_0 = 0 \\ cx_0 + dy_0 = 0. \end{cases}$$

If $a \neq 0$, we know that the first equation is satisfied if and only if

$$x_0 = -\frac{b}{a}y_0.$$

Now we see that any point that lies on this line $x_0 = (-b/a)y_0$ also satisfies the second linear equation $cx_0 + dy_0 = 0$. In fact, if we substitute a point of this form into the second component of the vector field, we have

$$cx_0 + dy_0 = c\left(-\frac{b}{a}\right)y_0 + dy_0$$

$$= \left(-\frac{bc}{a} + d\right)y_0$$

$$= \left(\frac{ad - bc}{a}\right)y_0$$

$$= \frac{\det A}{a}y_0$$

$$= 0,$$

since we are assuming that $\det A = 0$. Hence, the line $x_0 = (-b/a)y_0$ consists entirely of equilibrium points.

If $a = 0$ and $b \neq 0$, then the determinant condition $\det A = ad - bc = 0$ implies that $c = 0$. Consequently, the vector field at the point (x_0, y_0) is (by_0, dy_0). Since $b \neq 0$, we see that we get equilibrium points if and only if $y_0 = 0$. In other words, the set of equilibrium points is exactly the x-axis.

Finally, if $a = b = 0$, then the vector field at the point (x_0, y_0) is $(0, cx_0 + dy_0)$. In this case, we see that a point (x_0, y_0) is an equilibrium point if and only if $cx_0 + dy_0 = 0$. Since at least one of c or d is nonzero, the set of points (x_0, y_0) that satisfy $cx_0 + dy_0 = 0$ is precisely a line through the origin.

17. The first-order system corresponding to this equation is

$$\frac{dy}{dt} = v$$

$$\frac{dv}{dt} = -qy - pv.$$

(a) If $q = 0$, then the system becomes

$$\frac{dy}{dt} = v$$

$$\frac{dv}{dt} = -pv,$$

and the equilibrium points are the solutions of the system of equations

$$\begin{cases} v = 0 \\ -pv = 0. \end{cases}$$

Thus, the point (y, v) is an equilibrium point if and only if $v = 0$. In other words, the set of all equilibria agrees with the horizontal axis in the yv-plane.

(b) If $p = q = 0$, then the system becomes

$$\frac{dy}{dt} = v$$

$$\frac{dv}{dt} = 0$$

but the equilibrium points are again the points with $v = 0$.

19. Letting $v = dy/dt$ and $w = d^2y/dt^2$ we can write this equation as the system

$$\frac{dy}{dt} = v$$

$$\frac{dv}{dt} = \frac{d^2y}{dt^2} = w$$

$$\frac{dw}{dt} = \frac{d^3y}{dt^3} = -ry - pv - qw.$$

In matrix notation, this system is

$$\frac{d\mathbf{Y}}{dt} = \mathbf{AY}$$

where

$$\mathbf{Y} = \begin{pmatrix} y \\ v \\ w \end{pmatrix} \quad \text{and} \quad \mathbf{A} = \begin{pmatrix} 0 & 1 & 0 \\ 0 & 0 & 1 \\ -r & -p & -q \end{pmatrix}.$$

21. If there are fewer than the usual number of buyers, then $b < 0$. If this level of b has a negative effect on the number of sellers, we expect the γb-term in ds/dt to be negative. If $\gamma b < 0$ and $b < 0$, then we must have $\gamma > 0$.

23. In the model, $ds/dt = \gamma b + \delta s$. If $s > 0$, then the number of sellers is greater than usual and house prices should decline. Since declining prices should have a negative effect on the number of sellers, we expect the δs-term to be negative. If $\delta s < 0$ when $s > 0$, we should have $\delta < 0$.

25.   **(a)** Note that substituting $\mathbf{Y}(t)$ into the left-hand side of the differential equation, we get

$$\frac{d\mathbf{Y}}{dt} = \begin{pmatrix} e^{2t} + 2te^{2t} \\ -e^{2t} - 2(t+1)e^{2t} \end{pmatrix}$$

$$= \begin{pmatrix} e^{2t} + 2te^{2t} \\ -3e^{2t} - 2te^{2t} \end{pmatrix}.$$

Substituting $\mathbf{Y}(t)$ into the right-hand side, we get

$$\begin{pmatrix} 1 & -1 \\ 1 & 3 \end{pmatrix} \begin{pmatrix} te^{2t} \\ -(t+1)e^{2t} \end{pmatrix} = \begin{pmatrix} te^{2t} + (t+1)e^{2t} \\ te^{2t} - 3(t+1)e^{2t} \end{pmatrix}$$

$$= \begin{pmatrix} e^{2t} + 2te^{2t} \\ -3e^{2t} - 2te^{2t} \end{pmatrix}.$$

Since the left-hand side of the differential equation equals the right-hand side, the function $\mathbf{Y}(t)$ is a solution.

(b) At $t = 0$, $\mathbf{Y}(0) = (0, -1)$. By the Linearity Principle, any constant multiple of the solution $\mathbf{Y}(t)$ is also a solution. Since the function $-2\mathbf{Y}(t)$ has the desired initial condition, we know that

$$-2\mathbf{Y}(t) = \begin{pmatrix} -2te^{2t} \\ (2t+2)e^{2t} \end{pmatrix}$$

is the desired solution. By the Uniqueness Theorem, this is the only solution with this initial condition. (Given the formula for $-2\mathbf{Y}(t)$ directly above, note that we can directly check our assertion that this function solves the initial-value problem without appealing to the Linearity Principle.)

27.   **(a)** Substitute $\mathbf{Y}_1(t)$ into the differential equation and compare the left-hand side to the right-hand side. On the left-hand side, we have

$$\frac{d\mathbf{Y}_1}{dt} = \begin{pmatrix} -3e^{-3t} + 8e^{-4t} \\ -3e^{-3t} + 16e^{-4t} \end{pmatrix},$$

and on the right-hand side, we have

$$\mathbf{AY}_1(t) = \begin{pmatrix} -2 & -1 \\ 2 & -5 \end{pmatrix} \begin{pmatrix} e^{-3t} - 2e^{-4t} \\ e^{-3t} - 4e^{-4t} \end{pmatrix} = \begin{pmatrix} -3e^{-3t} + 8e^{-4t} \\ -3e^{-3t} + 16e^{-4t} \end{pmatrix}.$$

Since the two sides agree, we know that $Y_1(t)$ is a solution.
For $Y_2(t)$,

$$\frac{dY_2}{dt} = \begin{pmatrix} -6e^{-3t} - 4e^{-4t} \\ -6e^{-3t} - 8e^{-4t} \end{pmatrix},$$

and

$$AY_2(t) = \begin{pmatrix} -2 & -1 \\ 2 & -5 \end{pmatrix} \begin{pmatrix} 2e^{-3t} + e - 4t \\ 2e^{-3t} + 2e^{-4t} \end{pmatrix} = \begin{pmatrix} -6e^{-3t} - 4e^{-4t} \\ -6e^{-3t} - 8e^{-4t} \end{pmatrix}.$$

Since the two sides agree, the function $Y_2(t)$ is also a solution.

Both $Y_1(t)$ and $Y_2(t)$ are solutions, and we proceed to the next part of the exercise.

(b) Note that $Y_1(0) = (-1, -3)$ and $Y_2(0) = (3, 4)$. These vectors are not on the same line through the origin, so the initial conditions are linearly independent. If the initial conditions are linearly independent, then the solutions must also be linearly independent. Since the two solutions are linearly independent, we proceed to part (c) of the exercise.

(c) We must find constants k_1 and k_2 such that

$$k_1 Y_1(0) + k_2 Y_2(0) = k_1 \begin{pmatrix} -1 \\ -3 \end{pmatrix} + k_2 \begin{pmatrix} 3 \\ 4 \end{pmatrix} = \begin{pmatrix} 2 \\ 3 \end{pmatrix}.$$

In other words, the constants k_1 and k_2 must satisfy the simultaneous system of linear equations

$$\begin{cases} -k_1 + 3k_2 = 2 \\ -3k_1 + 4k_2 = 3. \end{cases}$$

It follows that $k_1 = -1/5$ and $k_2 = 3/5$. Hence, the required solution is

$$-\frac{1}{5}Y_1(t) + \frac{3}{5}Y_2(t) = \begin{pmatrix} e^{-3t} + e^{-4t} \\ e^{-3t} + 2e^{-4t} \end{pmatrix}.$$

29. (a) First, we check to see if $Y_1(t)$ is a solution. The left-hand side of the differential equation is

$$\frac{dY_1}{dt} = \begin{pmatrix} e^{-t} + 36e^{3t} \\ -e^{-t} + 12e^{3t} \end{pmatrix},$$

and the right-hand side is

$$AY_1(t) = \begin{pmatrix} 2 & 3 \\ 1 & 0 \end{pmatrix} \begin{pmatrix} -e^{-t} + 12e^{3t} \\ e^{-t} + 4e^{3t} \end{pmatrix} = \begin{pmatrix} e^{-t}36e^{3t} \\ -e^{-t} + 12e^{3t} \end{pmatrix}.$$

Consequently, $Y_1(t)$ is a solution. However,

$$\frac{dY_2}{dt} = \begin{pmatrix} e^{-t} \\ -2e^{-t} \end{pmatrix},$$

and

$$AY_2(t) = \begin{pmatrix} 2 & 3 \\ 1 & 0 \end{pmatrix} \begin{pmatrix} -e^{-t} \\ 2e^{-t} \end{pmatrix} = \begin{pmatrix} 4e^{-t} \\ -e^{-t} \end{pmatrix}.$$

Consequently, the function $Y_2(t)$ is not a solution. In this case, we are not able to solve the given initial-value problem, so we stop here.

31. **(a)** If $(x_1, y_1) = (0, 0)$, then (x_1, y_1) and (x_2, y_2) are on the same line through the origin because (x_1, y_1) is the origin. So (x_1, y_1) and (x_2, y_2) are linearly dependent.

(b) If $(x_1, y_1) = \lambda(x_2, y_2)$ for some λ, then (x_1, y_1) and (x_2, y_2) are on the same line through the origin. To see why, suppose that $x_2 \neq 0$ and $\lambda \neq 0$. (The $\lambda = 0$ case was handled in part (a) above.) In this case, $x_1 \neq 0$ as well. Then the slope of the line through the origin and (x_1, y_1) is y_1/x_1, and the slope of the line through the origin and (x_2, y_2) is y_2/x_2. However, because $(x_1, y_1) = \lambda(x_2, y_2)$, we have

$$\frac{y_1}{x_1} = \frac{\lambda y_2}{\lambda x_2} = \frac{y_2}{x_2}.$$

Since these two lines have the same slope and both contain the origin, they are the same line. (The special case where $x_2 = 0$ reduces to considering vertical lines through the origin.)

(c) If $x_1 y_2 - x_2 y_1 = 0$, then $x_1 y_2 = x_2 y_1$. Once again, this condition implies that (x_1, y_1) and (x_2, y_2) are on the same line through the origin. For example, suppose that $x_1 \neq 0$, then

$$y_2 = \frac{x_2 y_1}{x_1} = \frac{x_2}{x_1} y_1.$$

But we already know that

$$x_2 = \frac{x_2}{x_1} x_1,$$

so we have

$$(x_2, y_2) = \frac{x_2}{x_1}(x_1, y_1).$$

By part (b) above (where $\lambda = x_2/x_1$), the two vectors are linearly dependent.

If $x_1 = 0$ but $y_1 \neq 0$, it follows that $x_2 y_1 = 0$, and thus $x_2 = 0$. Thus, both (x_1, y_1) and (x_2, y_2) are on the vertical line through the origin.

Finally, if $x_1 = 0$ and $y_1 = 0$, we can use part (a) to show that the two vectors are linearly dependent.

33. The initial position of $Y_1(t)$ is $Y_1(0) = (-1, 1)$. By the Linearity Principle, we know that $kY_1(t)$ is also a solution of the system for any constant k. Hence, for any initial condition of the form $(-k, k)$, the solution is $kY_1(t)$.

(a) The curve $2Y_1(t) = (-2e^{-t}, 2e^{-t})$ is the solution with this initial condition.

(b) We cannot find the solution for this initial condition using only $Y_1(t)$.

(c) The constant function $0Y_1(t) = (0, 0)$ (represented by the equilibrium point at the origin) is the solution with this initial condition.

(d) The curve $-3Y_1(t) = (3e^{-t}, -3e^{-t})$ is the solution with this initial condition.

35. **(a)** Using the Product Rule we compute

$$\frac{dW}{dt} = \frac{dx_1}{dt}y_2 + x_1\frac{dy_2}{dt} - \frac{dx_2}{dt}y_1 - x_2\frac{dy_1}{dt}.$$

(b) Since $(x_1(t), y_1(t))$ and $(x_2(t), y_2(t))$ are solutions, we know that

$$\frac{dx_1}{dt} = ax_1 + by_1$$

$$\frac{dy_1}{dt} = cx_1 + dy_1$$

and that

$$\frac{dx_2}{dt} = ax_2 + by_2$$

$$\frac{dy_2}{dt} = cx_2 + dy_2.$$

Substituting these equations into the expression for dW/dt, we obtain

$$\frac{dW}{dt} = (ax_1 + by_1)y_2 + x_1(cx_2 + dy_2) - (ax_2 + by_2)y_1 - x_2(cx_1 + dy_1).$$

After we collect terms, we have

$$\frac{dW}{dt} = (a + d)W.$$

(c) This equation is a homogeneous, linear, first-order equation (as such it is also separable—see Sections 1.1, 1.2, and 1.8). Therefore, we know that the general solution is

$$W(t) = Ce^{(a+d)t}$$

where C is any constant (but note that $C = W(0)$).

(d) From Exercises 31 and 32, we know that $\mathbf{Y}_1(t)$ and $\mathbf{Y}_2(t)$ are linearly independent if and only if $W(t) \neq 0$. But, $W(t) = Ce^{(a+d)t}$, so $W(t) = 0$ if and only if $C = W(0) = 0$. Hence, $W(t) = 0$ is zero for some t if and only if $C = W(0) = 0$.

EXERCISES FOR SECTION 3.2

1. (a) The characteristic polynomial is

$$(3 - \lambda)(-2 - \lambda) = 0,$$

and therefore the eigenvalues are $\lambda_1 = -2$ and $\lambda_2 = 3$.

(b) To obtain the eigenvectors (x_1, y_1) for the eigenvalue $\lambda_1 = -2$, we solve the system of equations

$$\begin{cases} 3x_1 + 2y_1 = -2x_1 \\ -2y_1 = -2y_1 \end{cases}$$

and obtain $5x_1 = -2y_1$.

Using the same procedure, we see that the eigenvectors (x_2, y_2) for $\lambda_2 = 3$ must satisfy the equation $y_2 = 0$.

(c)

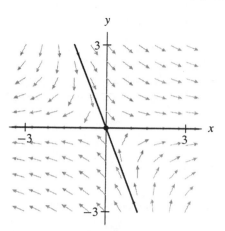

(d) One eigenvector $\mathbf{V}_1$ for λ_1 is $\mathbf{V}_1 = (-2, 5)$, and one eigenvector $\mathbf{V}_2$ for λ_2 is $\mathbf{V}_2 = (1, 0)$. Given the eigenvalues and these eigenvectors, we have the two linearly independent solutions

$$\mathbf{Y}_1(t) = e^{-2t} \begin{pmatrix} -2 \\ 5 \end{pmatrix} \quad \text{and} \quad \mathbf{Y}_2(t) = e^{3t} \begin{pmatrix} 1 \\ 0 \end{pmatrix}.$$

The $x(t)$- and $y(t)$-graphs for $\mathbf{Y}_1(t)$ are

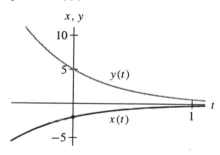

and the $x(t)$- and $y(t)$-graphs for $\mathbf{Y}_2(t)$ are

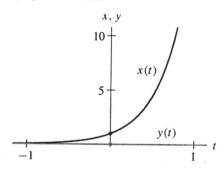

(e) The general solution to this linear system is

$$\mathbf{Y}(t) = k_1 e^{-2t} \begin{pmatrix} -2 \\ 5 \end{pmatrix} + k_2 e^{3t} \begin{pmatrix} 1 \\ 0 \end{pmatrix}.$$

3. **(a)** The characteristic polynomial is

$$(4 - \lambda)(3 - \lambda) - 2 = \lambda^2 - 7\lambda + 10 = 0,$$

and therefore the eigenvalues are $\lambda_1 = 2$ and $\lambda_2 = 5$.

(b) To obtain the eigenvectors (x_1, y_1) for the eigenvalue $\lambda_1 = 2$, we solve the system of equations

$$\begin{cases} 4x_1 + 2y_1 = 2x_1 \\ x_1 + 3y_1 = 2y_1 \end{cases}$$

and obtain $y_1 = -x_1$.

Using the same procedure, we obtain the eigenvectors (x_2, y_2) where $x_2 = 2y_2$ for $\lambda_2 = 5$.

(c)

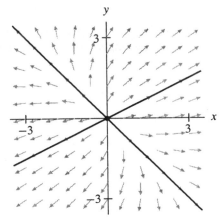

(d) One eigenvector $\mathbf{V}_1$ for λ_1 is $\mathbf{V}_1 = (1, -1)$, and one eigenvector $\mathbf{V}_2$ for λ_2 is $\mathbf{V}_2 = (2, 1)$. Given the eigenvalues and these eigenvectors, we have two linearly independent solutions

$$\mathbf{Y}_1(t) = e^{2t} \begin{pmatrix} 1 \\ -1 \end{pmatrix} \quad \text{and} \quad \mathbf{Y}_2(t) = e^{5t} \begin{pmatrix} 2 \\ 1 \end{pmatrix}.$$

The $x(t)$- and $y(t)$-graphs for $\mathbf{Y}_1(t)$ are

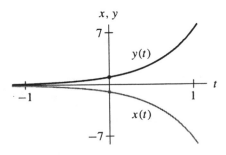

and the $x(t)$- and $y(t)$-graphs for $\mathbf{Y}_2(t)$ are

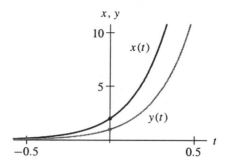

(e) The general solution to this linear system is

$$\mathbf{Y}(t) = k_1 e^{2t} \begin{pmatrix} 1 \\ -1 \end{pmatrix} + k_2 e^{5t} \begin{pmatrix} 2 \\ 1 \end{pmatrix}.$$

5. (a) The characteristic polynomial is

$$\left(-\tfrac{1}{2} - \lambda\right)^2 = 0,$$

and therefore there is only one eigenvalue, $\lambda = -1/2$.

(b) To obtain the eigenvectors (x_1, y_1) for the eigenvalue $\lambda = -1/2$, we solve the system of equations

$$\begin{cases} -\tfrac{1}{2}x_1 = -\tfrac{1}{2}x_1 \\ x_1 - \tfrac{1}{2}y_1 = -\tfrac{1}{2}y_1 \end{cases}$$

and obtain $x_1 = 0$.

(c)

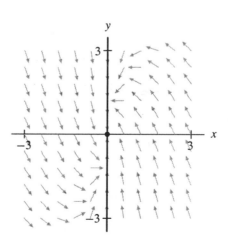

(d) Given the eigenvalue $\lambda = -1/2$ and the eigenvector $\mathbf{V} = (0, 1)$, we have the solution

$$\mathbf{Y}(t) = e^{-t/2} \begin{pmatrix} 0 \\ 1 \end{pmatrix}.$$

The $x(t)$- and $y(t)$-graphs for $\mathbf{Y}(t)$ are

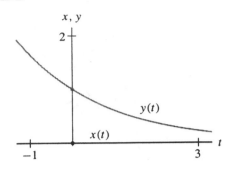

(e) Since the method of eigenvalues and eigenvectors does not give us a second solution that is linearly independent from $\mathbf{Y}(t)$, we cannot form the general solution.

7. (a) The characteristic polynomial is

$$(3 - \lambda)(-\lambda) - 4 = \lambda^2 - 3\lambda - 4 = (\lambda - 4)(\lambda + 1) = 0,$$

and therefore the eigenvalues are $\lambda_1 = -1$ and $\lambda_2 = 4$.

(b) To obtain the eigenvectors (x_1, y_1) for the eigenvalue $\lambda_1 = -1$, we solve the system of equations

$$\begin{cases} 3x_1 + 4y_1 = -x_1 \\ x_1 = -y_1 \end{cases}$$

and obtain $y_1 = -x_1$.

Using the same procedure, we obtain the eigenvectors (x_2, y_2) where $x_2 = 4y_2$ for $\lambda_2 = 4$.

(c)

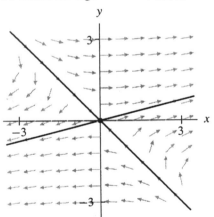

(d) One eigenvector $\mathbf{V}_1$ for λ_1 is $\mathbf{V}_1 = (1, -1)$, and one eigenvector $\mathbf{V}_2$ for λ_2 is $\mathbf{V}_2 = (4, 1)$. Given the eigenvalues and these eigenvectors, we have two linearly independent solutions

$$\mathbf{Y}_1(t) = e^{-t} \begin{pmatrix} 1 \\ -1 \end{pmatrix} \quad \text{and} \quad \mathbf{Y}_2(t) = e^{4t} \begin{pmatrix} 4 \\ 1 \end{pmatrix}.$$

The $x(t)$- and $y(t)$-graphs for $\mathbf{Y}_1(t)$ are

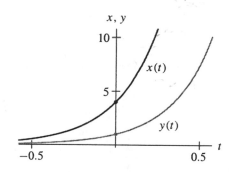

and the $x(t)$- and $y(t)$-graphs for $\mathbf{Y}_2(t)$ are

(e) The general solution to this linear system is

$$\mathbf{Y}(t) = k_1 e^{-t} \begin{pmatrix} 1 \\ -1 \end{pmatrix} + k_2 e^{4t} \begin{pmatrix} 4 \\ 1 \end{pmatrix}$$

9. (a) The characteristic polynomial is

$$(2 - \lambda)(1 - \lambda) - 1 = \lambda^2 - 3\lambda + 1 = 0,$$

and therefore the eigenvalues are

$$\lambda_1 = \frac{3 + \sqrt{5}}{2} \quad \text{and} \quad \lambda_2 = \frac{3 - \sqrt{5}}{2}.$$

(b) To obtain the eigenvectors (x_1, y_1) for the eigenvalue $\lambda_1 = (3 + \sqrt{5})/2$, we solve the system of equations

$$\begin{cases} 2x_1 + y_1 = \dfrac{3 + \sqrt{5}}{2} x_1 \\[2mm] x_1 + y_1 = \dfrac{3 + \sqrt{5}}{2} y_1 \end{cases}$$

and obtain

$$y_1 = \frac{-1 + \sqrt{5}}{2} x_1,$$

which is equivalent to the equation $2y_1 = (-1 + \sqrt{5})x_1$.

Using the same procedure, we obtain the eigenvectors (x_2, y_2) where $2y_2 = (-1 - \sqrt{5})x_2$ for $\lambda_2 = (3 - \sqrt{5})/2$.

(c)

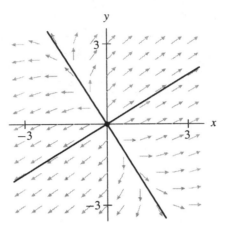

(d) One eigenvector $\mathbf{V}_1$ for the eigenvalue λ_1 is $\mathbf{V}_1 = (2, -1 + \sqrt{5})$, and one eigenvector $\mathbf{V}_2$ for the eigenvalue λ_2 is $\mathbf{V}_2 = (-2, 1 + \sqrt{5})$.

Given the eigenvalues and these eigenvectors, we have two linearly independent solutions

$$\mathbf{Y}_1(t) = e^{(3+\sqrt{5})t/2} \begin{pmatrix} 2 \\ -1 + \sqrt{5} \end{pmatrix} \quad \text{and} \quad \mathbf{Y}_2(t) = e^{(3-\sqrt{5})t/2} \begin{pmatrix} -2 \\ 1 + \sqrt{5} \end{pmatrix}.$$

The $x(t)$- and $y(t)$-graphs for $\mathbf{Y}_1(t)$ are

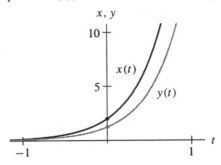

and the $x(t)$- and $y(t)$-graphs for $\mathbf{Y}_2(t)$ are

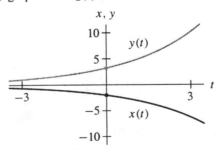

(e) The general solution to this linear system is

$$\mathbf{Y}(t) = k_1 e^{(3+\sqrt{5})t/2} \begin{pmatrix} 2 \\ -1 + \sqrt{5} \end{pmatrix} + k_2 e^{(3-\sqrt{5})t/2} \begin{pmatrix} -2 \\ 1 + \sqrt{5} \end{pmatrix}.$$

11. The characteristic polynomial is

$$(-3 - \lambda)(2 - \lambda) = 0,$$

and therefore the eigenvalues are $\lambda_1 = -3$ and $\lambda_2 = 2$.

To obtain the eigenvectors (x_1, y_1) for the eigenvalue $\lambda_1 = -3$, we solve the system of equations

$$\begin{cases} -3x_1 = -3x_1 \\ -x_1 + 2y_1 = -3y_1 \end{cases}$$

and obtain

$$5y_1 = x_1.$$

Therefore, an eigenvector for the eigenvalue $\lambda_1 = -3$ is $\mathbf{V}_1 = (5, 1)$.

Using the same procedure, we obtain the eigenvector $\mathbf{V}_2 = (0, 1)$ for $\lambda_2 = 2$.

Given the eigenvalues and these eigenvectors, we have two linearly independent solutions

$$\mathbf{Y}_1(t) = e^{-3t} \begin{pmatrix} 5 \\ 1 \end{pmatrix} \quad \text{and} \quad \mathbf{Y}_2(t) = e^{2t} \begin{pmatrix} 0 \\ 1 \end{pmatrix}.$$

The general solution to this linear system is

$$\mathbf{Y}(t) = k_1 e^{-3t} \begin{pmatrix} 5 \\ 1 \end{pmatrix} + k_2 e^{2t} \begin{pmatrix} 0 \\ 1 \end{pmatrix}.$$

(a) We have $\mathbf{Y}(0) = (1, 0)$, so we must find k_1 and k_2 so that

$$\begin{pmatrix} 1 \\ 0 \end{pmatrix} = \mathbf{Y}(0) = k_1 \begin{pmatrix} 5 \\ 1 \end{pmatrix} + k_2 \begin{pmatrix} 0 \\ 1 \end{pmatrix}.$$

This vector equation is equivalent to the simultaneous system of linear equations

$$\begin{cases} 5k_1 = 1 \\ k_1 + k_2 = 0. \end{cases}$$

Solving these equations, we obtain $k_1 = 1/5$ and $k_2 = -1/5$. Thus, the particular solution is

$$\mathbf{Y}(t) = \tfrac{1}{5} e^{-3t} \begin{pmatrix} 5 \\ 1 \end{pmatrix} - \tfrac{1}{5} e^{2t} \begin{pmatrix} 0 \\ 1 \end{pmatrix}.$$

Thus, $x(t) = e^{-3t}$ and $y(t) = (e^{-3t} - e^{2t})/5$.

(b) We have $\mathbf{Y}(0) = (0, 1)$. Since this initial condition is an eigenvector associated to the $\lambda = 2$ eigenvalue, we do not need to do any additional calculation. The desired solution to the initial-value problem is

$$\mathbf{Y}(t) = e^{2t} \begin{pmatrix} 0 \\ 1 \end{pmatrix}.$$

Thus, $x(t) = 0$ and $y(t) = e^{2t}$.

(c) We have $\mathbf{Y}(0) = (-2, 1)$, so we must find k_1 and k_2 so that

$$\begin{pmatrix} -2 \\ 1 \end{pmatrix} = \mathbf{Y}(0) = k_1 \begin{pmatrix} 5 \\ 1 \end{pmatrix} + k_2 \begin{pmatrix} 0 \\ 1 \end{pmatrix}.$$

This vector equation is equivalent to the simultaneous system of linear equations

$$\begin{cases} 5k_1 = -2 \\ k_1 + k_2 = 1. \end{cases}$$

Solving these equations, we obtain $k_1 = -2/5$ and $k_2 = 7/5$. Thus, the particular solution is

$$\mathbf{Y}(t) = -\tfrac{2}{5}e^{-3t} \begin{pmatrix} 5 \\ 1 \end{pmatrix} + \tfrac{7}{5}e^{2t} \begin{pmatrix} 0 \\ 1 \end{pmatrix}.$$

Thus, $x(t) = -2e^{-3t}$ and $y(t) = (-2e^{-3t} + 7e^{2t})/5$.

13. The characteristic polynomial is

$$(-4 - \lambda)(-3 - \lambda) - 2 = \lambda^2 + 7\lambda + 10 = 0,$$

and therefore the eigenvalues are $\lambda_1 = -5$ and $\lambda_2 = -2$.
   To obtain the eigenvectors (x_1, y_1) for the eigenvalue $\lambda_1 = -5$, we solve the system of equations

$$\begin{cases} -4x_1 + y_1 = -5x_1 \\ 2x_1 - 3y_1 = -5y_1 \end{cases}$$

and obtain

$$y_1 = -x_1.$$

Therefore, an eigenvector for the eigenvalue $\lambda_1 = -5$ is $\mathbf{V}_1 = (1, -1)$.
   Using the same procedure, we obtain the eigenvector $\mathbf{V}_2 = (1, 2)$ for $\lambda_2 = -2$.
   Given the eigenvalues and these eigenvectors, we have two linearly independent solutions

$$\mathbf{Y}_1(t) = e^{-5t} \begin{pmatrix} 1 \\ -1 \end{pmatrix} \quad \text{and} \quad \mathbf{Y}_2(t) = e^{-2t} \begin{pmatrix} 1 \\ 2 \end{pmatrix}.$$

The general solution to this linear system is

$$\mathbf{Y}(t) = k_1 e^{-5t} \begin{pmatrix} 1 \\ -1 \end{pmatrix} + k_2 e^{-2t} \begin{pmatrix} 1 \\ 2 \end{pmatrix}.$$

(a) We have $\mathbf{Y}(0) = (1, 0)$, so we must find k_1 and k_2 so that

$$\begin{pmatrix} 1 \\ 0 \end{pmatrix} = \mathbf{Y}(0) = k_1 \begin{pmatrix} 1 \\ -1 \end{pmatrix} + k_2 \begin{pmatrix} 1 \\ 2 \end{pmatrix}.$$

This vector equation is equivalent to the simultaneous system of linear equations

$$\begin{cases} k_1 + k_2 = 1 \\ -k_1 + 2k_2 = 0. \end{cases}$$

Solving these equations, we obtain $k_1 = 2/3$ and $k_2 = 1/3$. Thus, the particular solution is

$$\mathbf{Y}(t) = \tfrac{2}{3}e^{-5t}\begin{pmatrix} 1 \\ -1 \end{pmatrix} + \tfrac{1}{3}e^{-2t}\begin{pmatrix} 1 \\ 2 \end{pmatrix}.$$

(b) We have $\mathbf{Y}(0) = (2, 1)$, so we must find k_1 and k_2 so that

$$\begin{pmatrix} 2 \\ 1 \end{pmatrix} = \mathbf{Y}(0) = k_1\begin{pmatrix} 1 \\ -1 \end{pmatrix} + k_2\begin{pmatrix} 1 \\ 2 \end{pmatrix}.$$

This vector equation is equivalent to the simultaneous system of linear equations

$$\begin{cases} k_1 + k_2 = 2 \\ -k_1 + 2k_2 = 1. \end{cases}$$

Solving these equations, we obtain $k_1 = 1$ and $k_2 = 1$. Thus, the particular solution is

$$\mathbf{Y}(t) = e^{-5t}\begin{pmatrix} 1 \\ -1 \end{pmatrix} + e^{-2t}\begin{pmatrix} 1 \\ 2 \end{pmatrix}.$$

(c) We have $\mathbf{Y}(0) = (-1, -2)$. Since this initial condition is an eigenvector associated to the $\lambda = -2$ eigenvalue, we do not need to do any additional calculation. The desired solution to the initial-value problem is

$$\mathbf{Y}(t) = -e^{-2t}\begin{pmatrix} 1 \\ 2 \end{pmatrix}.$$

15. Given any vector $\mathbf{Y}_0 = (x_0, y_0)$, we have

$$\mathbf{A}\mathbf{Y}_0 = \begin{pmatrix} a & 0 \\ 0 & a \end{pmatrix}\begin{pmatrix} x_0 \\ y_0 \end{pmatrix} = \begin{pmatrix} ax_0 \\ ay_0 \end{pmatrix} = a\begin{pmatrix} x_0 \\ y_0 \end{pmatrix} = a\mathbf{Y}_0.$$

Therefore, every nonzero vector is an eigenvector associated to the eigenvalue a.

17. The characteristic polynomial of $\mathbf{B}$ is

$$\lambda^2 - (a + d)\lambda + ad - b^2.$$

The roots of this polynomial are

$$\begin{aligned} \lambda &= \frac{a + d \pm \sqrt{(a + d)^2 - 4(ad - b^2)}}{2} \\ &= \frac{a + d \pm \sqrt{a^2 + 2ad + d^2 - 4ad + 4b^2}}{2} \\ &= \frac{a + d \pm \sqrt{(a - d)^2 + 4b^2}}{2}. \end{aligned}$$

Since the discriminant $D = (a - d)^2 + 4b^2$ is always nonnegative, the roots λ are real. Therefore, the matrix $\mathbf{B}$ has real eigenvalues. If $b \neq 0$, then D is positive and hence $\mathbf{B}$ has two distinct eigenvalues. (The only way to have only one eigenvalue is for $D = 0$).

19.   (a) To form the system, we introduce the new dependent variable $v = dy/dt$. Then

$$\frac{dv}{dt} = \frac{d^2y}{dt^2} = -p\frac{dy}{dt} - qy = -pv - qy.$$

Written in matrix form this system where $\mathbf{Y} = (y, v)$, we have

$$\frac{d\mathbf{Y}}{dt} = \begin{pmatrix} 0 & 1 \\ -q & -p \end{pmatrix} \mathbf{Y}.$$

(b) The characteristic polynomial is

$$(0 - \lambda)(-p - \lambda) + q = \lambda^2 + p\lambda + q.$$

(c) The roots of this polynomial (the eigenvalues) are

$$\frac{-p \pm \sqrt{p^2 - 4q}}{2}.$$

(d) The roots are distinct real numbers if the discriminant $D = p^2 - 4q$ is positive. In other words, the roots are distinct real numbers if $p^2 > 4q$.

(e) Since q is positive, $p^2 - 4q < p^2$, so we know that $\sqrt{p^2 - 4q} < \sqrt{p^2} = p$. Since the numerator in the expression for the eigenvalues is $-p \pm \sqrt{p^2 - 4q}$, we see that it must be negative. Since the denominator is positive, the eigenvalues must be negative.

21.   (a) Given $v = dy/dt$, the corresponding system is

$$\frac{dy}{dt} = v$$

$$\frac{dv}{dt} = 10y - 3v.$$

(b) The characteristic polynomial is $\lambda^2 + 3\lambda - 10 = (\lambda + 5)(\lambda - 2) = 0$, and the eigenvalues are $\lambda_1 = -5$ and $\lambda_2 = 2$.
To find the eigenvectors $\mathbf{V}_1 = (y_1, v_1)$ associated to the eigenvalue $\lambda_1 = -5$, we solve the system of equations

$$\begin{cases} v_1 = -5y_1 \\ 10y_1 - 3v_1 = -5v_1 \end{cases}$$

and obtain $v_1 = -5y_1$.
By the same procedure, we can find the eigenvectors $\mathbf{V}_2 = (y_2, v_2)$ for the eigenvalue $\lambda_2 = 2$. They consist of all vectors that satisfy the equation $v_2 = 2y_2$.

(c) From part (b) we see that one eigenvector for $\lambda_1 = -5$ is $\mathbf{V}_1 = (1, -5)$. Therefore the solution $\mathbf{Y}_1(t)$ that satisfies $\mathbf{Y}_1(0) = \mathbf{V}_1$ is

$$\mathbf{Y}_1(t) = e^{-5t} \begin{pmatrix} 1 \\ -5 \end{pmatrix}.$$

One eigenvector for $\lambda_2 = 2$ is $V_2 = (1, 2)$, and the solution $Y_2(t)$ that satisfies $Y_2(0) = V_2$ is

$$Y_2(t) = e^{2t} \begin{pmatrix} 1 \\ 2 \end{pmatrix}.$$

(d) Note that the solutions obtained here are vector-valued functions of the form $Y(t) = (y(t), v(t))$. In Section 2.3 we obtained $y_1(t) = e^{2t}$ and $y_2(t) = e^{-5t}$. Using the fact that $v = dy/dt$, we can obtain $Y_1(t)$ and $Y_2(t)$ from $y_1(t)$ and $y_2(t)$.

23. **(a)** Given $v = dy/dt$, the corresponding system is

$$\frac{dy}{dt} = v$$

$$\frac{dv}{dt} = -y - 4v.$$

(b) The characteristic polynomial is $\lambda^2 + 4\lambda + 1 = 0$. Using the quadratic formula, we obtain the eigenvalues $\lambda_1 = -2 + \sqrt{3}$ and $\lambda_2 = -2 - \sqrt{3}$.

To find the eigenvectors $V_1 = (y_1, v_1)$ associated to the eigenvalue $\lambda_1 = -2 + \sqrt{3}$, we solve the system of equations

$$\begin{cases} v_1 = (-2 + \sqrt{3})y_1 \\ -y_1 - 4v_1 = (-2 + \sqrt{3})v_1 \end{cases}$$

and obtain $v_1 = (-2 + \sqrt{3})y_1$.

By the same procedure, we can find the eigenvectors $V_2 = (y_2, v_2)$ for the eigenvalue $\lambda_2 = -2 - \sqrt{3}$. They consist of all vectors that satisfy the equation $v_2 = (-2 - \sqrt{3})y_2$.

(c) From part (b) we see that one eigenvector for $\lambda_1 = -2 + \sqrt{3}$ is $V_1 = (1, -2 + \sqrt{3})$. Therefore the solution $Y_1(t)$ that satisfies $Y_1(0) = V_1$ is

$$Y_1(t) = e^{(-2+\sqrt{3})t} \begin{pmatrix} 1 \\ -2 + \sqrt{3} \end{pmatrix}.$$

One eigenvector for $\lambda_2 = -2 + \sqrt{3}$ is $V_2 = (1, -2 - \sqrt{3})$, and the solution $Y_2(t)$ that satisfies $Y_2(0) = V_2$ is

$$Y_2(t) = e^{(-2-\sqrt{3})t} \begin{pmatrix} 1 \\ -2 - \sqrt{3} \end{pmatrix}.$$

(d) Note that the solutions obtained here are vector-valued functions of the form $Y(t) = (y(t), v(t))$. In Section 2.3 we obtained $y_1(t) = e^{(-2+\sqrt{3})t}$ and $y_2(t) = e^{(-2-\sqrt{3})t}$. Using the fact that $v = dy/dt$, we can obtain $Y_1(t)$ and $Y_2(t)$ from $y_1(t)$ and $y_2(t)$.

25. With $m = 1$, $k = 4$, and $b = 1$, the system is

$$\frac{dy}{dt} = v$$

$$\frac{dv}{dt} = -4y - v.$$

The characteristic polynomial is $\lambda^2 + \lambda + 4$, and its roots are the complex numbers $(-1 \pm \sqrt{15}\,i)/2$. Therefore there are no straight-line solutions. According to the direction field, the solution curves seem to spiral around the origin.

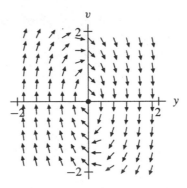

EXERCISES FOR SECTION 3.3

1. As we computed in Exercise 1 of Section 3.2, the eigenvalues are $\lambda_1 = -2$ and $\lambda_2 = 3$. The eigenvectors (x_1, y_1) for the eigenvalue $\lambda_1 = -2$ satisfy $5x_1 = -2y_1$, and the eigenvectors (x_2, y_2) for $\lambda_2 = 3$ satisfy the equation $y_2 = 0$. The equilibrium point at the origin is a saddle.

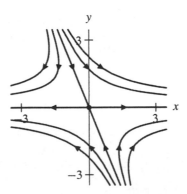

3. As we computed in Exercise 3 of Section 3.2, the eigenvalues are $\lambda_1 = 2$ and $\lambda_2 = 5$. The eigenvectors (x_1, y_1) for the eigenvalue $\lambda_1 = 2$ satisfy $y_1 = -x_1$, and the eigenvectors (x_2, y_2) for $\lambda_2 = 5$ satisfy $x_2 = 2y_2$. The equilibrium point at the origin is a source.

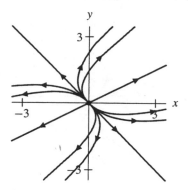

5. As we computed in Exercise 7 of Section 3.2, the eigenvalues are $\lambda_1 = -1$ and $\lambda_2 = 4$. The eigenvectors (x_1, y_1) for the eigenvalue $\lambda_1 = -1$ satisfy $y_1 = -x_1$, and the eigenvectors (x_2, y_2) for $\lambda_2 = 4$ satisfy $x_2 = 4y_2$. The equilibrium point at the origin is a saddle.

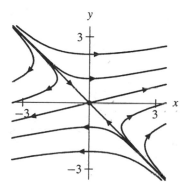

7. As we computed in Exercise 9 of Section 3.2, the eigenvalues are

$$\lambda_1 = \frac{3 + \sqrt{5}}{2} \quad \text{and} \quad \lambda_2 = \frac{3 - \sqrt{5}}{2}.$$

The eigenvectors (x_1, y_1) for the eigenvalue $\lambda_1 = (3 + \sqrt{5})/2$ satisfy

$$y_1 = \frac{-1 + \sqrt{5}}{2} x_1,$$

and the eigenvectors (x_2, y_2) for $\lambda_2 = (3 - \sqrt{5})/2$ satisfy

$$y_2 = \frac{-1 - \sqrt{5}}{2} x_2.$$

The equilibrium point at the origin is a source.

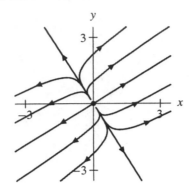

9. As we computed in Exercise 11 of Section 3.2, the eigenvalues are $\lambda_1 = -3$ and $\lambda_2 = 2$. The eigenvectors (x_1, y_1) for the eigenvalue $\lambda_1 = -3$ satisfy $5y_1 = x_1$, and the eigenvectors (x_2, y_2) for the eigenvalue $\lambda_2 = 2$ satisfy $x_2 = 0$. The equilibrium point at the origin is a saddle. Therefore, the solution curves in the phase plane for the initial conditions $(1, 0)$, $(0, 1)$, and $(-2, 1)$ are shown in the following figure.

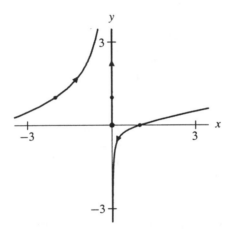

(a) The solution curve with initial condition $(1, 0)$ is asymptotic to the negative y-axis as $t \to \infty$ and is asymptotic to the line $y = x/5$ in the first quadrant as $t \to -\infty$.

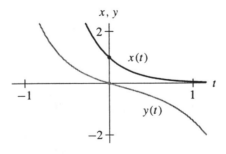

(b) The solution curve with initial condition $(0, 1)$ lies entirely on the positive y-axis, and $y(t) \to \infty$ in an exponential fashion as $t \to \infty$.

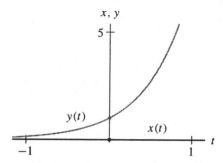

(c) The solution curve with initial condition $(-2, 1)$ is asymptotic to the positive y-axis as $t \to \infty$ and is asymptotic to the line $y = x/5$ in the third quadrant as $t \to -\infty$.

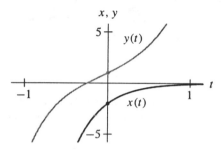

11. As we computed in Exercise 13 of Section 3.2, the eigenvalues are $\lambda_1 = -5$ and $\lambda_2 = -2$. The eigenvectors (x_1, y_1) for the eigenvalue $\lambda_1 = -5$ satisfy $y_1 = -x_1$, and the eigenvectors (x_2, y_2) for the eigenvalue $\lambda_2 = -2$ satisfy $y_2 = 2x_2$. The equilibrium point at the origin is a sink. The solution curves in the phase plane for the initial conditions $(1, 0)$, $(2, 1)$, and $(-1, -2)$ are shown in the following figure.

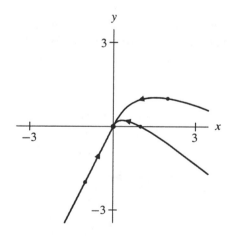

(a) The solution curve with initial condition $(1, 0)$ approaches the origin tangent to the line $y = 2x$.

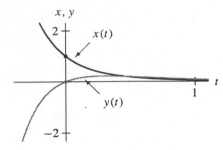

(b) The solution curve with initial condition $(2, 1)$ approaches the origin tangent to the line $y = 2x$.

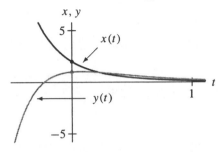

(c) The initial condition $(-1, -2)$ is an eigenvector associated to the eigenvalue $\lambda_2 = -2$. The corresponding solution curve approaches the origin along the line $y = 2x$ as $t \to \infty$.

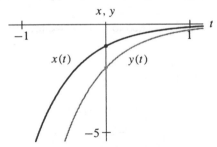

13. As we computed in Exercise 21 of Section 3.2, the eigenvalues are $\lambda_1 = -5$ and $\lambda_2 = 2$. The eigenvectors (y_1, v_1) associated to the eigenvalue $\lambda_1 = -5$ satisfy $v_1 = -5y_1$, and the eigenvectors (y_2, v_2) for the eigenvalue $\lambda_2 = 2$ satisfy the equation $v_2 = 2y_2$. The equilibrium point at the origin is a saddle.

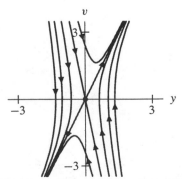

15. As we computed in Exercise 23 of Section 3.2, the eigenvalues are $\lambda_1 = -2+\sqrt{3}$ and $\lambda_2 = -2-\sqrt{3}$. The eigenvectors (y_1, v_1) associated to the eigenvalue $\lambda_1 = -2 + \sqrt{3}$ satisfy the equation $v_1 = (-2 + \sqrt{3})y_1$, and the eigenvectors (y_2, v_2) for the eigenvalue $\lambda_2 = -2 - \sqrt{3}$ satisfy the equation $v_2 = (-2 - \sqrt{3})y_2$. The equilibrium point at the origin is a sink.

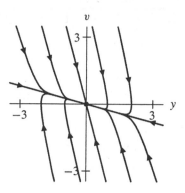

17. The characteristic equation is

$$(2 - \lambda)(-1 - \lambda) = 0,$$

and therefore, the eigenvalues are $\lambda_1 = 2$ and $\lambda_2 = -1$. The equilibrium point at the origin is a saddle.

To compute the eigenvectors associated to $\lambda_1 = 2$, we must solve the simultaneous equations

$$\begin{cases} 2x + y = 2x \\ -y = 2y. \end{cases}$$

Therefore, any vector of the form $(x, 0)$ is an eigenvector associated to the eigenvalue $\lambda_1 = 2$.

Similarly, for $\lambda_1 = -1$, the eigenvectors (x, y) associated to the eigenvalue $\lambda_2 = -1$ must satisfy the equation $y = -3x$. Therefore, we know that the phase portrait is

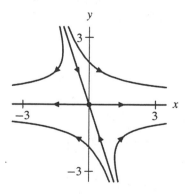

Given an initial condition on the line $y = -3x$, the corresponding solution is a straight-line solution that is asymptotic to the origin. For any other initial condition, the corresponding solution is

asymptotic to the x-axis. Therefore, for any initial condition, Bob's profits, $y(t)$, eventually tend to 0 as $t \to \infty$.

To see what happens to Paul's profits, we must locate the initial condition relative to the line $y = -3x$. As stated above, if the initial condition (x_0, y_0) lies on the line $y = -3x$, then Paul's profits will also tend to 0 eventually. However, if (x_0, y_0) lies to the left of the line $y = -3x$, Paul goes broke. On the other hand, if (x_0, y_0) lies to the right of the line $y = -3x$, then Paul makes a fortune.

19. **(a)** The characteristic equation is
$$(-2 - \lambda)(-1 - \lambda) = 0,$$

so the eigenvalues are $\lambda_1 = -2$ and $\lambda_2 = -1$. Therefore, the equilibrium point at the origin is a sink.

(b) To find all the straight-line solutions, we must calculate the eigenvectors. For the eigenvalue $\lambda_1 = -2$, we have the simultaneous equations

$$\begin{cases} -2x_1 + \frac{1}{2}y_1 = -2x_1 \\ \quad -y_1 = -2y_1. \end{cases}$$

The second equation implies that $y_1 = 0$. In other words, all vectors on the x-axis are eigenvectors for λ_1. Therefore, any solution of the form $e^{-2t}(x_1, 0)$ for any x_1 is a straight-line solution corresponding to the eigenvalue $\lambda_1 = -2$.

To calculate the eigenvectors associated to the eigenvalue $\lambda_2 = -1$, we must solve the equations

$$\begin{cases} -2x_2 + \frac{1}{2}y_2 = -x_2 \\ \quad -y_2 = -y_2. \end{cases}$$

From the first equation, we see that $y_2 = 2x_2$. Therefore, any solution of the form $e^{-t}(x_2, 2x_2)$ for any x_2 is a straight-line solution corresponding to the eigenvalue $\lambda_2 = -1$.

(c) In the phase plane, all solution curves approach the origin as $t \to \infty$. If the initial condition is on the x-axis, it yields a straight-line solution that remains on the x-axis as $t \to \infty$. For any other initial condition, the solution approaches the origin tangent to the line $y = 2x$.

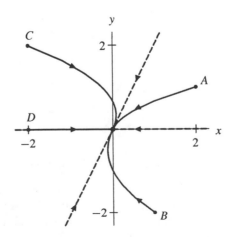

For the initial condition $A = (2, 1)$, the solution curve remains in the first quadrant. Since it approaches the origin tangent to the line $y = 2x$, it must cross the line $y = x$ at some time. Therefore, the $x(t)$- and $y(t)$-graphs are positive for all t, but they cross at some time $t > 0$.

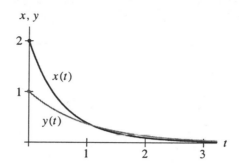

For the initial condition $B = (1, -2)$, we see that $y(t)$ is increasing but negative for all t. We also see that $x(t)$ is decreasing initially. It becomes negative, reaches a minimum, and then increases as it approaches 0. Note that $x(t) \neq y(t)$ for all t.

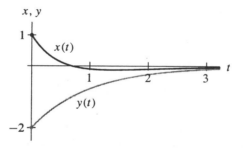

For the initial condition $C = (-2, 2)$, we see that $y(t)$ is decreasing but positive for all t. We also see that $x(t)$ is increasing initially. It becomes positive, reaches a maximum, and then decreases as it approaches 0. Again, these two graphs do not cross at any time.

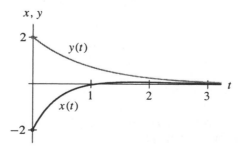

The initial condition $D = (-2, 0)$ lies on the line of eigenvectors associated to the eigenvalue $\lambda_1 = -1$. Therefore, the solution curve remains on the x-axis for all t. Hence, $y(t) = 0$ for all t, and $x(t)$ is the exponential function $-2e^{-2t}$.

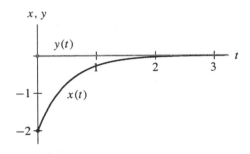

21. **(a)** The second-order equation is

$$\frac{d^2y}{dt^2} + 7\frac{dy}{dt} + 6y = 0.$$

Introducing $v = dy/dt$, we obtain the system

$$\frac{dy}{dt} = v$$

$$\frac{dv}{dt} = -6y - 7v.$$

(b) The characteristic polynomial is

$$\lambda^2 + 7\lambda + 6,$$

which factors into $(\lambda + 6)(\lambda + 1)$.

(c) From the characteristic polynomial, we obtain the eigenvalues $\lambda_1 = -6$ and $\lambda_2 = -1$.

(d) To compute the eigenvectors associated to $\lambda_1 = -6$, we solve the simultaneous equations

$$\begin{cases} v = -6y \\ -6y - 7v = -6v. \end{cases}$$

Therefore, any vector on the line $v = -6y$ is an eigenvector associated to the eigenvalue λ_1. To compute the eigenvectors associated to $\lambda_2 = -1$, we must solve the simultaneous equations

$$\begin{cases} v = -y \\ -6y - 7v = -v. \end{cases}$$

Therefore, any vector on the line $y = -v$ is an eigenvector associated to the eigenvalue λ_2. Since both eigenvalues are real and negative, we know that origin is a sink, and the solution curve corresponding to the initial condition $(y(0), v(0)) = (2, 0)$ tends toward the origin tangent to the line $y = -v$ in the yv-plane.

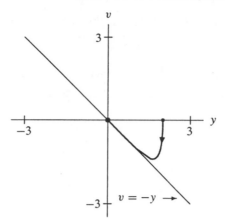

From the phase portrait, we see that the solution curve remains in the fourth quadrant for all $t > 0$. Consequently, it does not cross the line $y = 0$, and the mass cannot cross the equilibrium position. The solution approaches the origin at the rate that is determined by the eigenvalue $\lambda_2 = -1$. In other words, it approaches the origin at the rate of e^{-t}.

23. **(a)** Written in terms of its components, the system is

$$\frac{dx}{dt} = -0.2x - 0.1y$$

$$\frac{dy}{dt} = -0.1y.$$

Since the coefficient of y in dx/dt is negative, the introduction of new fish ($y > 0$) contributes negatively to dx/dt. Hence, the new fish have a negative affect on the native fish population. Since the equation for dy/dt involves only y, the native fish have no affect on the population of the new fish.

(b) If the new species of fish is not introduced (that is, if $y(t) = 0$ for all t), then the system reduces to $dx/dt = -0.2x$. In this case, we have an exponential decay model as in Section 1.1, and the native fish population tends to its equilibrium level. (Remember: the quantity $x(t)$ is the difference between the native fish population and its equilibrium level, not the actual of fish population.) Thus, the model agrees with the system as described.

(c) There are two lines consisting of straight-line solutions, and the solutions with initial conditions on these lines are asymptotic to the origin as $t \rightarrow 0$. To find these lines, we must compute the eigenvalues and eigenvectors.
The characteristic polynomial of the given matrix is

$$(-0.2 - \lambda)(-0.1 - \lambda),$$

and hence the eigenvalues are $\lambda_1 = -0.2$ and $\lambda_2 = -0.1$.
To find an eigenvector for $\lambda_1 = -0.2$, we must solve

$$\begin{pmatrix} -0.2 & -0.1 \\ 0.0 & -0.1 \end{pmatrix} \begin{pmatrix} x_0 \\ y_0 \end{pmatrix} = -0.2 \begin{pmatrix} x_0 \\ y_0 \end{pmatrix}.$$

Rewritten in terms of components, this equation becomes

$$\begin{cases} -0.2x_0 - 0.1y_0 = -0.2x_0 \\ \qquad\quad -0.1y_0 = -0.2y_0, \end{cases}$$

which is equivalent to

$$\begin{cases} -0.1y_0 = 0 \\ \ \ \ 0.1y_0 = 0. \end{cases}$$

If we multiply the second equation by -1, we obtain the first equation. Therefore, the equations are redundant and any vector (x_0, y_0) that satisfies the first equation is an eigenvector. Setting $x_0 = 1$ yields the eigenvector $\mathbf{V}_1 = (1, 0)$.

To find an eigenvector for $\lambda_2 = -0.1$, we repeat the process with λ_2 in replace of λ_1, and we obtain the eigenvector $\mathbf{V}_2 = (1, -1)$

Since both eigenvalues are negative, solutions with initial conditions that lie on the lines through the origin determined by the eigenvectors (the x-axis and the line $y = -x$) tend toward the origin.

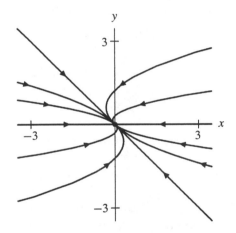

(d) Using the phase portrait shown in part (c), we see that solutions with initial conditions of the form $(0, y)$, $y > 0$, move through the second quadrant and tend toward the equilibrium point at the origin. Thus, our model predicts that, if a small number of new fish are added to the lake, the native population drops below its equilibrium level since x is negative. The new fish will die out and the native fish will return to equilibrium.

25. **(a)** Written in terms of its components, the system is

$$\frac{dx}{dt} = -0.2x + 0.1y$$

$$\frac{dy}{dt} = -0.1y.$$

Since the dy/dt equation does not depend on x, the native fish have no affect on the population of the new fish. Since the coefficient of y in the equation for dx/dt is positive, the introduction of new fish increases the population of the native fish.

(b) If the new species of fish is not introduced (that is, if $y(t) = 0$ for all t), then the above system simply becomes $dx/dt = -0.2x$. In this case, we have an exponential decay model as in Section 1.1, and the native fish population tends to its equilibrium level. (Remember: the quantity $x(t)$ is the difference between the native fish population and its equilibrium level, not the actual of fish population.) Thus, the model agrees with the system as described.

(c) The characteristic equation is

$$(-0.2 - \lambda)(-0.1 - \lambda) = 0,$$

and the eigenvalues are $\lambda_1 = -0.2$ and $\lambda_2 = -0.1$.
To find an eigenvector for $\lambda_1 = -0.2$, we must solve the simultaneous equations

$$\begin{cases} -0.2x + 0.1y = -0.2x \\ -0.1y = -0.2y. \end{cases}$$

Therefore, $y = 0$. One such eigenvector $\mathbf{V}_1$ is $(1, 0)$.
For $\lambda = -0.1$, the simultaneous equations are

$$\begin{cases} -0.2x + 0.1y = -0.1x \\ -0.1y = -0.1y. \end{cases}$$

Any vector that satisfies $y = x$ satisfies these equations. One such eigenvector is $\mathbf{V}_2 = (1, 1)$.

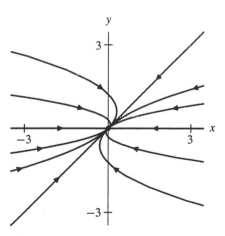

(d) Using the phase portrait shown in part (c), we see that solutions with initial conditions of the form $(0, y)$, $y > 0$, move through the first quadrant and tend toward the equilibrium point at the origin. Thus, our model predicts that, if a small number of new fish are added to the lake, the population of native fish increases above its equilibrium value and the population of new fish decreases. Eventually, the new fish head toward extinction, and the native fish return to their equilibrium population.

27. **(a)** The characteristic equation is

$$(-2 - \lambda)(2 - \lambda) = 0.$$

Therefore, the eigenvalues are $\lambda_1 = -2$ and $\lambda_2 = 2$, and the equilibrium point at the origin is a saddle.

(b) To find the eigenvectors (x_1, y_1) corresponding to $\lambda_1 = -2$, we solve the simultaneous linear equations

$$\begin{cases} -2x + y = -2x \\ 2y = -2y. \end{cases}$$

Therefore, the eigenvectors lie on the line $y = 0$, the x-axis. Similarly, the eigenvectors associated to the eigenvalue $\lambda_2 = 2$ lie on the line $y = 4x$.

Using these eigenvalues and eigenvectors, we can give a rough sketch of the phase portrait.

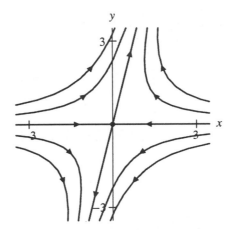

(c)

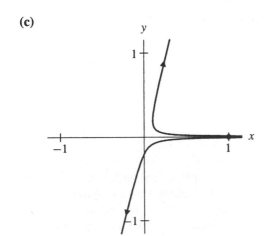

(d) The eigenvalues are -2 and 2 with eigenvectors $(1, 0)$ and $(1, 4)$ respectively, The initial conditions are on either side of the line of eigenvectors corresponding to eigenvalue -2. Hence, these two solutions will approach the origin along the x-axis, but the y coordinates will grow, at approximately the rate e^{2t}. Since the initial separation is 0.02 and we seek the approximate time t when the separation is 1, we must solve $0.02e^{2t} = 1$, which yields $t = \ln(50)/2 \approx 1.96$.

EXERCISES FOR SECTION 3.4

1. Using Euler's formula, we can write

$$e^{(1+3i)t} \begin{pmatrix} 2+i \\ 1 \end{pmatrix} = e^t e^{3it} \begin{pmatrix} 2+i \\ 1 \end{pmatrix}$$

$$= e^t (\cos 3t + i \sin 3t) \begin{pmatrix} 2+i \\ 1 \end{pmatrix}$$

$$= e^t \begin{pmatrix} 2\cos 3t - \sin 3t \\ \cos 3t \end{pmatrix} + i e^t \begin{pmatrix} 2\sin 3t + \cos 3t \\ \sin 3t \end{pmatrix}.$$

Hence, we have

$$\mathbf{Y}_{re}(t) = e^t \begin{pmatrix} 2\cos 3t - \sin 3t \\ \cos 3t \end{pmatrix} \quad \text{and} \quad \mathbf{Y}_{im}(t) = e^t \begin{pmatrix} \cos 3t + 2\sin 3t \\ \sin 3t \end{pmatrix}.$$

3. **(a)** The characteristic equation is

$$(-\lambda)^2 + 4 = \lambda^2 + 4 = 0,$$

and the eigenvalues are $\lambda = \pm 2i$.

(b) Since the real part of the eigenvalues are 0, the origin is a center.

(c) Since $\lambda = \pm 2i$, the natural period is $2\pi/2 = \pi$, and the natural frequency is $1/\pi$.

(d) At $(1, 0)$, the tangent vector is $(-2, 0)$. Therefore, the direction of oscillation is clockwise.

(e) According to the phase plane, $x(t)$ and $y(t)$ are periodic with period π. At the initial condition $(1, 0)$, both $x(t)$ and $y(t)$ are initially decreasing.

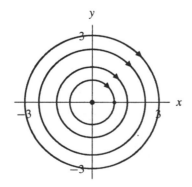

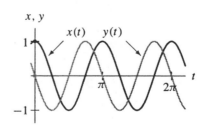

5. **(a)** The characteristic polynomial is

$$(-1 - \lambda)(-1 - \lambda) + 2 = \lambda^2 + 2\lambda + 3,$$

so the eigenvalues are $\lambda = -1 \pm i\sqrt{2}$.

(b) The eigenvalues are complex and the real part is negative so the origin is a spiral sink.

(c) The natural period is $2\pi/\sqrt{2} = \sqrt{2}\,\pi$. The natural frequency is $1/(\sqrt{2}\,\pi)$.

(d) At the point $(1, 0)$, the vector field is $(-1, -1)$. Hence, the solution curves must spiral in a clockwise fashion.

(e)

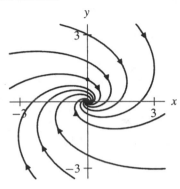
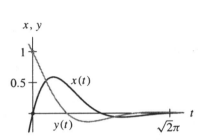

7. **(a)** The characteristic equation is

$$(2 - \lambda)(1 - \lambda) + 12 = \lambda^2 + 3\lambda + 14 = 0,$$

and the eigenvalues are $\lambda = (3 \pm \sqrt{47}\,i)/2$.

(b) Since the real part of the eigenvalues is positive, the origin is a spiral source.

(c) Since $\lambda = (3 \pm \sqrt{47}i)/2$, natural period is $4\pi/\sqrt{47}$, and natural frequency is $\sqrt{47}/(4\pi)$.

(d) At the point $(1, 0)$, the tangent vector is $(2, 2)$. Therefore, the solution curves spiral about the origin in a counterclockwise fashion.

(e) From the phase plane, we see that both $x(t)$ and $y(t)$ oscillate and that the amplitude of these oscillations increases rapidly.

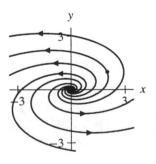
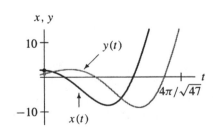

9. **(a)** According to Exercise 3, $\lambda = \pm 2i$. The eigenvectors (x, y) associated to eigenvalue $\lambda = 2i$ must satisfy the equation $2y = 2ix$, which is equivalent to $y = ix$. One such eigenvector is $(1, i)$, and thus we have the complex solution

$$\mathbf{Y}(t) = e^{2it} \begin{pmatrix} 1 \\ i \end{pmatrix} = \begin{pmatrix} \cos 2t \\ -\sin 2t \end{pmatrix} + i \begin{pmatrix} \sin 2t \\ \cos 2t \end{pmatrix}.$$

Taking real and imaginary parts, we obtain the general solution

$$\mathbf{Y}(t) = k_1 \begin{pmatrix} \cos 2t \\ -\sin 2t \end{pmatrix} + k_2 \begin{pmatrix} \sin 2t \\ \cos 2t \end{pmatrix}.$$

(b) From the initial condition, we obtain

$$k_1 \begin{pmatrix} 1 \\ 0 \end{pmatrix} + k_2 \begin{pmatrix} 0 \\ 1 \end{pmatrix} = \begin{pmatrix} 1 \\ 0 \end{pmatrix},$$

and therefore, $k_1 = 1$ and $k_2 = 0$. Consequently, the solution with the initial condition $(1, 0)$ is

$$\mathbf{Y}(t) = \begin{pmatrix} \cos 2t \\ -\sin 2t \end{pmatrix}.$$

(c)

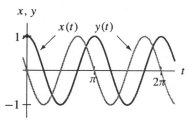

11. **(a)** To find the general solution, we find the eigenvectors from the characteristic polynomial

$$(-1 - \lambda)(-1 - \lambda) + 2 = \lambda^2 + 2\lambda + 3.$$

The eigenvalues are $\lambda = -1 \pm i\sqrt{2}$. To find an eigenvector associated to the eigenvector $-1 + i\sqrt{2}$, we must solve the equations

$$\begin{cases} -x + 2y = \left(-1 + i\sqrt{2}\right) x \\ -x - y = \left(-1 + i\sqrt{2}\right) y. \end{cases}$$

We see that the eigenvectors must satisfy the equation $2y = i\sqrt{2}\,x$. Using the eigenvector $(2, i\sqrt{2})$, we obtain the complex-valued solution

$$\mathbf{Y}(t) = e^{\left(-1 + i\sqrt{2}\right)t} \begin{pmatrix} 2 \\ i\sqrt{2} \end{pmatrix}.$$

Using Euler's formula, we write $\mathbf{Y}(t)$ as

$$\mathbf{Y}(t) = e^{-t}\left(\cos\sqrt{2}\,t + i\sin\sqrt{2}\,t\right)\begin{pmatrix} 2 \\ i\sqrt{2} \end{pmatrix},$$

which can be expressed as

$$\mathbf{Y}(t) = e^{-t}\begin{pmatrix} 2\cos\sqrt{2}\,t \\ -\sqrt{2}\sin\sqrt{2}\,t \end{pmatrix} + ie^{-t}\begin{pmatrix} 2\sin\sqrt{2}\,t \\ \sqrt{2}\cos\sqrt{2}\,t \end{pmatrix}.$$

Taking real and imaginary parts, we can form the general solution

$$k_1e^{-t}\begin{pmatrix} 2\cos\sqrt{2}\,t \\ -\sqrt{2}\sin\sqrt{2}\,t \end{pmatrix} + k_2e^{-t}\begin{pmatrix} 2\sin\sqrt{2}\,t \\ \sqrt{2}\cos\sqrt{2}\,t \end{pmatrix}.$$

(b) To find the particular solution with initial condition $(0, 1)$, we solve for k_1 and k_2 and obtain

$$\begin{cases} 2k_1 = 0 \\ \sqrt{2}k_2 = 1. \end{cases}$$

We have $k_1 = 0$ and $k_2 = 1/\sqrt{2}$.
The desired solution is

$$\mathbf{Y}(t) = e^{-t}\begin{pmatrix} \sqrt{2}\sin\sqrt{2}\,t \\ \cos\sqrt{2}\,t \end{pmatrix}.$$

(c)

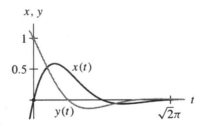

13. **(a)** According to Exercise 7, the eigenvalues are $\lambda = (3 \pm i\sqrt{47})/2$. The eigenvalues (x, y) associated to the eigenvector $(3 + i\sqrt{47})/2$ must satisfy the equation $12y = (1 - i\sqrt{47})x$. Hence, one eigenvector is $(12, 1 - i\sqrt{47})$, and we have the complex-valued solution

$$\mathbf{Y}(t) = e^{(3+i\sqrt{47})t/2}\begin{pmatrix} 12 \\ 1 - i\sqrt{47} \end{pmatrix}$$

$$= e^{3t/2}\begin{pmatrix} 12\cos\left(\frac{\sqrt{47}}{2}t\right) \\ \cos\left(\frac{\sqrt{47}}{2}t\right) + \sqrt{47}\sin\left(\frac{\sqrt{47}}{2}t\right) \end{pmatrix} + $$

$$ie^{3t/2}\begin{pmatrix} 12\sin\left(\frac{\sqrt{47}}{2}t\right) \\ -\sqrt{47}\cos\left(\frac{\sqrt{47}}{2}t\right) + \sin\left(\frac{\sqrt{47}}{2}t\right) \end{pmatrix}.$$

Taking real and imaginary parts

$$\mathbf{Y}_{\text{re}}(t) = e^{3t/2} \begin{pmatrix} 12\cos\left(\frac{\sqrt{47}}{2}t\right) \\ \cos\left(\frac{\sqrt{47}}{2}t\right) + \sqrt{47}\sin\left(\frac{\sqrt{47}}{2}t\right) \end{pmatrix}$$

and

$$\mathbf{Y}_{\text{im}}(t) = e^{3t/2} \begin{pmatrix} 12\sin\left(\frac{\sqrt{47}}{2}t\right) \\ -\sqrt{47}\cos\left(\frac{\sqrt{47}}{2}t\right) + \sin\left(\frac{\sqrt{47}}{2}t\right) \end{pmatrix},$$

we obtain the general solution $k_1\mathbf{Y}_{\text{re}}(t) + k_2\mathbf{Y}_{\text{im}}(t)$.

(b) From the initial condition, we have

$$k_1 \begin{pmatrix} 12 \\ 1 \end{pmatrix} + k_2 \begin{pmatrix} 0 \\ -\sqrt{47} \end{pmatrix} = \begin{pmatrix} 2 \\ 1 \end{pmatrix}.$$

Thus, $k_1 = 1/6$ and $k_2 = -5/(6\sqrt{47})$, and the desired solution is

$$\mathbf{Y}(t) = e^{3t/2} \begin{pmatrix} 2\cos\left(\frac{\sqrt{47}}{2}t\right) - \frac{10}{\sqrt{47}}\sin\left(\frac{\sqrt{47}}{2}t\right) \\ \cos\left(\frac{\sqrt{47}}{2}t\right) + \frac{7}{\sqrt{47}}\sin\left(\frac{\sqrt{47}}{2}t\right) \end{pmatrix}.$$

(c)

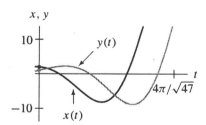

15. **(a)** In the case of complex eigenvalues, the function $x(t)$ oscillates about $x = 0$ with constant period, and the amplitude of successive oscillations is either increasing, decreasing, or constant depending on the sign of the real part of the eigenvalue. The graphs that satisfy these properties are (ii) and (v).

(b) For (ii), the natural period is about 1.5 and, since the amplitude tends toward zero as t increases, the origin is a sink. For (v), the natural period is about 1.25 and, since the amplitude increases as t increases, the origin is a source.

(c) (i) The time between successive zeros is not constant.
(iii) The amplitude is not monotonically decreasing or increasing.
(iv) Oscillation stops at some t.
(vi) Oscillation starts at some t. There was no prior oscillation.

17. We know that $\lambda_1 = \alpha + i\beta$ satisfies the equation $\lambda_1^2 + a\lambda_1 + b = 0$. Therefore, if we take the complex conjugate all of the terms in this equation, we obtain

$$(\alpha - i\beta)^2 + a(\alpha - i\beta) + b = 0,$$

since a and b are real. The complex conjugate of λ_1 is $\lambda_2 = \alpha - i\beta$, and we have

$$\lambda_2^2 + a\lambda_2 + b = 0.$$

Therefore, λ_2 is also a root.

19. Suppose $Y_2 = kY_1$ for some constant k. Then, $Y_0 = (1 + ik)Y_1$. Since $AY_0 = \lambda Y_0$, we have

$$(1 + ik)AY_1 = \lambda(1 + ik)Y_1.$$

Thus, $AY_1 = \lambda Y_1$. Now note that the left-hand side, AY_1, is a real vector. However, since λ is complex and Y_1 is real, the right-hand side is complex (that is, it has a nonzero imaginary part). Thus, we have a contradiction, and Y_1 and Y_2 must be linearly independent.

21. **(a)** The factor $e^{-\alpha t}$ is positive for all t. Hence, the zeros of $x(t)$ are exactly the zeros of $\sin \beta t$. Suppose t_1 and t_2 are successive zeros (that is, $t_1 < t_2$, $x(t_1) = x(t_2) = 0$, and $x(t) \neq 0$ for $t_1 < t < t_2$), then $\beta t_2 - \beta t_1 = \pi$. In other words, $t_2 - t_1 = \pi/\beta$.

(b) By the nature of sine function, local maxima and local minima appear alternately. Therefore, we look for t_1 and t_2 such that $x'(t_1) = x'(t_2) = 0$ and $x'(t) \neq 0$ for $t_1 < t < t_2$. From

$$x'(t) = e^{-\alpha t}(-\alpha \sin \beta t + \beta \cos \beta t) = 0,$$

we know that $\tan \beta t = \beta/\alpha$ if t is corresponds to a local extremum. Since the tangent function is periodic with period π, $\beta(t_2 - t_1) = \pi$. Hence, $t_2 - t_1 = \pi/\beta$. Note that the distance between a local minimum and the following local maximum of $x(t)$ is constant over t.

(c) From part (b), we know that the distance between the first local maximum and the first local minimum is π/β and the distance between the first local minimum and the second local maximum is π/β. Therefore, the distance between the first two local maxima of $x(t)$ is $2\pi/\beta$.

(d) From part (b), we know that the first local maximum of $x(t)$ occurs at $t = (\arctan(\beta/\alpha))/\beta$.

23. **(a)** The corresponding first-order system is

$$\frac{dy}{dt} = v$$

$$\frac{dv}{dt} = -qy - pv.$$

(b) The characteristic polynomial is

$$(-\lambda)(-p - \lambda) + q = \lambda^2 + p\lambda + q,$$

so the eigenvalues are $\lambda = (-p \pm \sqrt{p^2 - 4q})/2$. Hence, the eigenvalues are complex if and only if $p^2 < 4q$.

(c) In order to have a spiral sink, we must have $p^2 < 4q$ (to make the eigenvalues complex) and $p > 0$ (to make the real part of the eigenvalues negative). To have a center, we must have $p = 0$.

(d) The vector field at $(1, 0)$ is $(0, -q)$. Hence, if $q > 0$, then the vector field points down along the entire y-axis, and the solution curves spiral about the origin in a clockwise fashion.

25. There is no spiral saddle because a linear saddle is a linear system where some solutions approach the origin and some move away. If one solution spirals toward (or away from) the origin, then we can multiply that solution by any constant, scaling it so that it goes through any point in the plane. This scaled solution is still a solution of the system (recall the Linearity Principle), so every solution spirals in the same way, either toward or away from the origin.

EXERCISES FOR SECTION 3.5

1.   (a) The characteristic equation is

$$(-3 - \lambda)^2 = 0,$$

and the eigenvalue is $\lambda = -3$.

(b) To find an eigenvector, we solve the simultaneous equations

$$\begin{cases} -3x = -3x \\ x - 3y = -3y. \end{cases}$$

Then, $x = 0$, and one eigenvector is $(0, 1)$.

(c) Note the straight-line solutions along the y-axis.

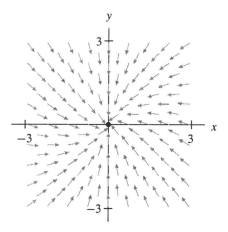

(d) Since the eigenvalue is negative, any solution with an initial condition on the y-axis tends toward the origin as t increases. According to the direction field, every solution tends to the origin as t increases. The solutions with initial conditions in the half-plane $x > 0$ eventually approach the origin along the positive y-axis. Similarly, the solutions with initial conditions in the half-plane $x < 0$ eventually approach the origin along the negative y-axis.

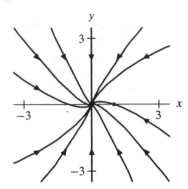

(e) At the point $Y_0 = (1, 0)$, $dY/dt = (-3, 1)$. Therefore, $x(t)$ decreases initially and $y(t)$ increases initially. The solution eventually approaches the origin tangent to the positive y-axis. Therefore, $x(t)$ monotonically decreases to zero and $y(t)$ eventually decreases toward zero. Since the solution with the initial condition Y_0 never crosses y-axis in the phase plane, the function $x(t) > 0$ for all t.

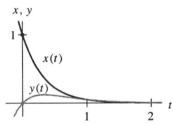

3. **(a)** The characteristic equation is

$$(-2 - \lambda)(-4 - \lambda) + 1 = (\lambda + 3)^2 = 0,$$

and the eigenvalue is $\lambda = -3$.

(b) To find an eigenvector, we solve the simultaneous equations

$$\begin{cases} -2x - y = -3x \\ x - 4y = -3y. \end{cases}$$

Then, $y = x$, and one eigenvector is $(1, 1)$.

(c) Note the straight-line solutions along the line $y = x$.

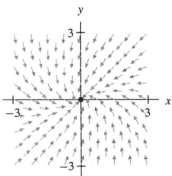

(d) Since the eigenvalue is negative, any solution on the line $y = x$ tends toward the origin along $y = x$ as t increases. According to the direction field, every solution tends to the origin as t increases. The solutions with initial conditions that lie in the half-plane $y > x$ eventually approach the origin tangent to the half-line $y = x$ with $y < 0$. Similarly, the solutions with initial conditions that lie in the half-plane $y < x$ eventually approach the origin tangent to the line $y = x$ with $y > 0$.

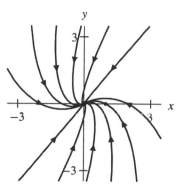

(e) At the point $\mathbf{Y}_0 = (1, 0)$, $d\mathbf{Y}/dt = (-2, 1)$. Therefore, $x(t)$ initially decreases and $y(t)$ initially increases. The solution eventually approaches the origin tangent to the line $y = x$. Since the solution curve never crosses the line $y = x$, the graphs of $x(t)$ and $y(t)$ do not cross.

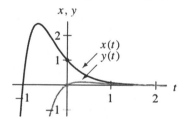

5. (a) According to Exercise 1, there is one eigenvalue, -3, with eigenvectors of the form $(0, y_0)$, where $y_0 \neq 0$.

To find the general solution, we start with an arbitrary initial condition $\mathbf{V}_0 = (x_0, y_0)$. Then

$$\mathbf{V}_1 = \left[\begin{pmatrix} -3 & 0 \\ 1 & -3 \end{pmatrix} + 3 \begin{pmatrix} 1 & 0 \\ 0 & 1 \end{pmatrix} \right] \mathbf{V}_0$$

$$= \begin{pmatrix} 0 & 0 \\ 1 & 0 \end{pmatrix} \begin{pmatrix} x_0 \\ y_0 \end{pmatrix}$$

$$= \begin{pmatrix} 0 \\ x_0 \end{pmatrix}.$$

We obtain the general solution

$$\mathbf{Y}(t) = e^{-3t} \begin{pmatrix} x_0 \\ y_0 \end{pmatrix} + t e^{-3t} \begin{pmatrix} 0 \\ x_0 \end{pmatrix}.$$

(b) The solution that satisfies the initial condition $(x_0, y_0) = (1, 0)$ is

$$\mathbf{Y}(t) = e^{-3t} \begin{pmatrix} 1 \\ 0 \end{pmatrix} + te^{-3t} \begin{pmatrix} 0 \\ 1 \end{pmatrix}.$$

Hence, $x(t) = e^{-3t}$ and $y(t) = te^{-3t}$.

(c) Compare the graphs of $x(t) = e^{-3t}$ and $y(t) = te^{-3t}$ with the sketches obtained in part (e) of Exercise 1.

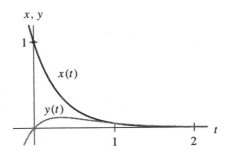

7. **(a)** From Exercise 3, we know that there is only one eigenvalue, $\lambda = -3$, and the eigenvectors (x_0, y_0) satisfy the equation $y_0 = x_0$.

To find the general solution, we start with an arbitrary initial condition $\mathbf{V}_0 = (x_0, y_0)$. Then

$$\mathbf{V}_1 = \left[\begin{pmatrix} -2 & -1 \\ 1 & -4 \end{pmatrix} + 3 \begin{pmatrix} 1 & 0 \\ 0 & 1 \end{pmatrix} \right] \mathbf{V}_0$$

$$= \begin{pmatrix} 1 & -1 \\ 1 & -1 \end{pmatrix} \begin{pmatrix} x_0 \\ y_0 \end{pmatrix}$$

$$= \begin{pmatrix} x_0 - y_0 \\ x_0 - y_0 \end{pmatrix}.$$

We obtain the general solution

$$\mathbf{Y}(t) = e^{-3t} \begin{pmatrix} x_0 \\ y_0 \end{pmatrix} + te^{-3t} \begin{pmatrix} x_0 - y_0 \\ x_0 - y_0 \end{pmatrix}.$$

(b) The solution that satisfies the initial condition $(x_0, y_0) = (1, 0)$ is

$$\mathbf{Y}(t) = e^{-3t} \begin{pmatrix} 1 \\ 0 \end{pmatrix} + te^{-3t} \begin{pmatrix} 1 \\ 1 \end{pmatrix}.$$

Hence, $x(t) = e^{-3t}(t + 1)$ and $y(t) = te^{-3t}$.

(c) Compare the graphs of $x(t) = e^{-3t}(t + 1)$ and $y(t) = te^{-3t}$ with the sketches obtained in part (e) of Exercise 3.

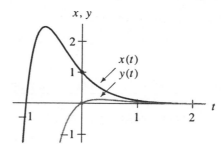

9. **(a)** By solving the quadratic equation, we obtain

$$\lambda = \frac{-\alpha \pm \sqrt{\alpha^2 - 4\beta}}{2}.$$

Therefore, for the quadratic to have a double root, we must have

$$\alpha^2 - 4\alpha\beta = 0.$$

(b) If zero is a root, we set $\lambda = 0$ in $\lambda^2 + \alpha\lambda + \beta = 0$, and we obtain $\beta = 0$.

11. The characteristic equation is

$$-\lambda(-p - \lambda) + q = \lambda^2 + p\lambda + q = 0.$$

Solving the quadratic equation, one obtains

$$\lambda = \frac{-p \pm \sqrt{p^2 - 4q}}{2}.$$

(a) Therefore, in order for **A** to have two real eigenvalues, p and q must satisfy $p^2 - 4q > 0$.
(b) In order for **A** to have complex eigenvalues, p and q must satisfy $p^2 - 4q < 0$.
(c) In order for **A** to have only one eigenvalue, p and q must satisfy $p^2 - 4q = 0$.

13. Since every vector is an eigenvector with eigenvalue λ, we substitute $\mathbf{Y} = (1, 0)$ into the equation $\mathbf{AY} = \lambda\mathbf{Y}$ and get

$$\mathbf{A}\begin{pmatrix} 1 \\ 0 \end{pmatrix} = \begin{pmatrix} a \\ c \end{pmatrix} = \lambda\begin{pmatrix} 1 \\ 0 \end{pmatrix}.$$

Hence, $a = \lambda$ and $c = 0$. Similarly, letting $\mathbf{Y} = (0, 1)$, we have

$$\begin{pmatrix} b \\ d \end{pmatrix} = \lambda\begin{pmatrix} 0 \\ 1 \end{pmatrix}.$$

Therefore, $b = 0$ and $d = \lambda$.

15. Since $\mathbf{Y}_1(0) = \mathbf{V}_0$ and $\mathbf{Y}_2(0) = \mathbf{W}_0$, we see that $\mathbf{V}_0 = \mathbf{W}_0$.
    Evaluating at $t = 1$ yields

$$\mathbf{Y}_1(1) = e^\lambda(\mathbf{V}_0 + \mathbf{V}_1) \quad \text{and} \quad \mathbf{Y}_2(1) = e^\lambda(\mathbf{W}_0 + \mathbf{W}_1).$$

Since $\mathbf{Y}_1(1) = \mathbf{Y}_2(1)$ and $\mathbf{V}_0 = \mathbf{W}_0$, we see that $\mathbf{V}_1 = \mathbf{W}_1$.

17. **(a)** The characteristic polynomial is

$$(-\lambda)(-1 - \lambda) + 0 = \lambda^2 + \lambda,$$

so the eigenvalues are $\lambda = 0$ and $\lambda = -1$.

(b) To find the eigenvectors $\mathbf{V}_1$ associated to the eigenvalue $\lambda = 0$, we must solve $\mathbf{AV}_1 = 0\mathbf{V}_1 = 0$ where $\mathbf{A}$ is the matrix that defines this linear system. (Note that this is the same calculation we do if we want to locate the equilibrium points.) We get

$$\begin{cases} 2y_1 = 0 \\ -y_1 = 0, \end{cases}$$

where $\mathbf{V}_1 = (x_1, y_1)$. Hence, the eigenvectors associated to $\lambda = 0$ (as well as the equilibrium points) must satisfy the equation $y_1 = 0$.

To find the eigenvectors $\mathbf{V}_2$ associated to the eigenvalue $\lambda = -1$, we must solve $\mathbf{AV}_2 = -\mathbf{V}_2$. We get

$$\begin{cases} 2y_2 = -x_2 \\ -y_2 = -y_2. \end{cases}$$

where $\mathbf{V}_2 = (x_2, y_2)$. Hence, the eigenvectors associated to $\lambda = -1$ must satisfy $2y_2 = -x_2$.

(c) The equation $y_1 = 0$ specifies a line of equilibrium points. Since the other eigenvalue is negative, solution curves not corresponding to equilibria move toward this line as t increases.

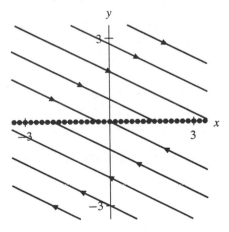

(d) Since $(1, 0)$ is an equilibrium point, it is easy to sketch the corresponding $x(t)$- and $y(t)$-graphs.

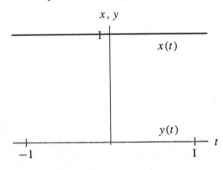

(e) To form the general solution, we must pick one eigenvector for each eigenvalue. Using part (b), we pick $V_1 = (1, 0)$, and $V_2 = (2, -1)$. We obtain the general solution

$$Y(t) = k_1 \begin{pmatrix} 1 \\ 0 \end{pmatrix} + k_2 e^{-t} \begin{pmatrix} 2 \\ -1 \end{pmatrix}.$$

(f) To determine the solution whose initial condition is $(1, 0)$, we can substitute $t = 0$ in the general solution and solve for k_1 and k_2. However, since this initial condition is an equilibrium point, we need not make the effort. We simply observe that

$$Y(t) = \begin{pmatrix} 1 \\ 0 \end{pmatrix}$$

is the desired solution.

19. (a) The characteristic polynomial is

$$(4 - \lambda)(1 - \lambda) - 4 = \lambda^2 - 5\lambda,$$

so the eigenvalues are $\lambda = 0$ and $\lambda = 5$.

(b) To find the eigenvectors V_1 associated to the eigenvalue $\lambda = 0$, we must solve $AV_1 = 0V_1 = 0$ where A is the matrix that defines this linear system. (Note that this is the same calculation we do if we want to locate the equilibrium points.) We get

$$\begin{cases} 4x_1 + 2y_1 = 0 \\ 2x_1 + y_1 = 0, \end{cases}$$

where $V_1 = (x_1, y_1)$. Hence, the eigenvectors associated to $\lambda = 0$ (as well as the equilibrium points) must satisfy the equation $y_1 = -2x_1$.

To find the eigenvectors V_2 associated to the eigenvalue $\lambda = 5$, we must solve $AV_2 = 5V_2$. We get

$$\begin{cases} 4x_2 + 2y_2 = 5x_2 \\ 2x_2 + y_2 = 5y_2. \end{cases}$$

where $V_2 = (x_2, y_2)$. Hence, the eigenvectors associated to $\lambda = 5$ must satisfy $x_2 = 2y_2$.

(c) The equation $y_1 = -2x_1$ specifies a line of equilibrium points. Since the other eigenvalue is positive, solution curves not corresponding to equilibria move away from this line as t increases.

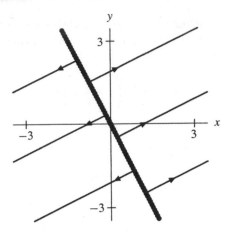

(d) As t increases, both $x(t)$ and $y(t)$ increase exponentially. As t decreases, both x and y approach constants that are determined by the line of equilibrium points.

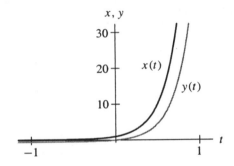

(e) To form the general solution, we must pick one eigenvector for each eigenvalue. Using part (b), we pick $\mathbf{V}_1 = (1, -2)$, and $\mathbf{V}_2 = (2, 1)$. We obtain the general solution

$$\mathbf{Y}(t) = k_1 \begin{pmatrix} 1 \\ -2 \end{pmatrix} + k_2 e^{5t} \begin{pmatrix} 2 \\ 1 \end{pmatrix}.$$

(f) To determine the solution whose initial condition is $(1, 0)$, we let $t = 0$ in the general solution and obtain the equations

$$k_1 \begin{pmatrix} 1 \\ -2 \end{pmatrix} + k_2 \begin{pmatrix} 2 \\ 1 \end{pmatrix} = \begin{pmatrix} 1 \\ 0 \end{pmatrix}.$$

Therefore, $k_1 = 1/5$ and $k_2 = 2/5$, and the particular solution is

$$\mathbf{Y}(t) = \begin{pmatrix} \frac{1}{5} + \frac{4}{5} e^{5t} \\ -\frac{2}{5} + \frac{2}{5} e^{5t} \end{pmatrix}.$$

21. **(a)** The characteristic polynomial is $\lambda^2 = 0$, so $\lambda = 0$ is the sole eigenvalue. To sketch the phase portrait we note that $dy/dt = 0$, so $y(t)$ is always a constant function. Moreover, $dx/dt = 2y$, so $x(t)$ is increasing if $y > 0$, and it is decreasing if $y < 0$.

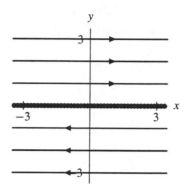

(b) This system is exactly the same as the one in part (a) except that the sign of dx/dt has changed. Hence, the phase portrait is the identical except for the fact that the arrows point the other way.

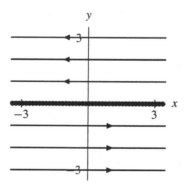

23. **(a)** The characteristic polynomial is $(a - \lambda)(d - \lambda)$, so the eigenvalues are a and d.

(b) If $a \neq d$, the lines of eigenvectors for a and d are the x- and y-axes respectively.

(c) If $a = d < 0$, every vector is an eigenvector (see Exercise 14), and all the vectors point toward the origin. Hence, every solution curve is asymptotic to the origin along a straight line.

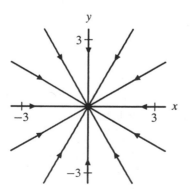

(d) The only difference between this case and part (c) is that the arrows in the vector field are reversed. Every solution tends away from the origin along a straight line.

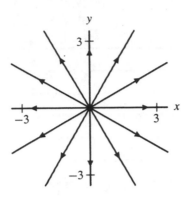

EXERCISES FOR SECTION 3.6

1. The characteristic polynomial is

$$s^2 + 3s - 10,$$

so the eigenvalues are $s = 2$ and $s = -5$. Hence, the general solution is

$$y(t) = k_1 e^{2t} + k_2 e^{-5t}.$$

3. The characteristic polynomial is

$$s^2 + 6s + 9,$$

so $s = -3$ is a repeated eigenvalue. Hence, the general solution is

$$y(t) = k_1 e^{-3t} + k_2 t e^{-3t}.$$

5. The characteristic polynomial is

$$s^2 + 8s + 25,$$

so the complex eigenvalues are $s = -4 \pm 3i$. Hence, the general solution is

$$y(t) = k_1 e^{-4t} \cos 3t + k_2 e^{-4t} \sin 3t.$$

7. The characteristic polynomial is

$$s^2 + 5s + 6,$$

so the eigenvalues are $s = -2$ and $s = -3$. Hence, the general solution is

$$y(t) = k_1 e^{-2t} + k_2 e^{-3t},$$

and we have

$$y'(t) = -2k_1 e^{-2t} - 3k_2 e^{-3t}.$$

From the initial conditions, we obtain the simultaneous equations

$$\begin{cases} k_1 + k_2 = 0 \\ -2k_1 - 3k_2 = 2. \end{cases}$$

Solving for k_1 and k_2 yields $k_1 = 2$ and $k_2 = -2$. Hence, the solution to our initial-value problem is $y(t) = 2e^{-2t} - 2e^{-3t}$.

9. The characteristic polynomial is

$$s^2 + 2s + 5,$$

so the eigenvalues are $s = -1 \pm 2i$. Hence, the general solution is

$$y(t) = k_1 e^{-t} \cos 3t + k_2 e^{-t} \sin 3t.$$

From the initial condition $y(0) = 3$, we see that $k_1 = 3$. Differentiating

$$y(t) = 3e^{-t} \cos 3t + k_2 e^{-t} \sin 3t$$

and evaluating $y'(t)$ at $t = 0$ yields $y'(0) = -3 + 2k_2$. Since $y'(0) = -1$, we have $k_2 = 1$. Hence, the solution to our initial-value problem is

$$y(t) = 3e^{-t} \cos 2t + e^{-t} \sin 2t.$$

11. The characteristic polynomial is

$$s^2 + 2s + 1,$$

so $s = -1$ is a repeated eigenvalue. Hence, the general solution is

$$y(t) = k_1 e^{-t} + k_2 t e^{-t}.$$

From the initial condition $y(0) = 1$, we see that $k_1 = 1$. Differentiating

$$y(t) = e^{-t} + k_2 t e^{-t}$$

and evaluating $y'(t)$ at $t = 0$ yields $y'(0) = -1 + k_2$. Since $y'(0) = 1$, we have $k_2 = 2$. Hence, the solution to our initial-value problem is

$$y(t) = e^{-t} + 2t e^{-t}.$$

13. **(a)** The resulting second-order equation is

$$\frac{d^2y}{dt^2} + 8\frac{dy}{dt} + 7y = 0,$$

and the corresponding system is

$$\frac{dy}{dt} = v$$
$$\frac{dv}{dt} = -7y - 8v.$$

(b) Recall that we can read off the characteristic equation of the second-order equation straight from the equation without having to revert to the corresponding system. We obtain

$$\lambda^2 + 8\lambda + 7 = 0.$$

Therefore, the eigenvalues are $\lambda_1 = -1$ and $\lambda_2 = -7$.
To find the eigenvectors associated to the eigenvalue λ_1, we solve the simultaneous system of equations

$$\begin{cases} v = -y \\ -7y - 8v = -v. \end{cases}$$

From the first equation, we immediately see that the eigenvectors associated to this eigenvalue must satisfy $v = -y$. Similarly, the eigenvectors associated to the eigenvalue $\lambda_2 = -7$ must satisfy the equation $v = -7y$.

(c) Since the eigenvalues are real and negative, the equilibrium point at the origin is a sink, and the system is overdamped.

(d) We know that all solution curves approach the origin as $t \to \infty$ and, with the exception of those whose initial conditions lie on the line $v = -7y$, these solution curves approach the origin tangent to the line $v = -y$.

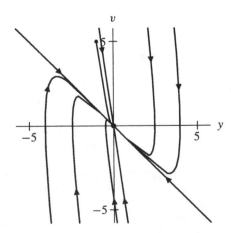

(e) From the phase portrait, we see that $y(t)$ increases monotonically toward 0 as $t \to \infty$. Also, $v(t)$ decreases monotonically toward 0. It is useful to remember that $v = dy/dt$.

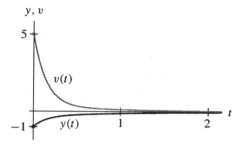

15. **(a)** The resulting second-order equation is

$$\frac{d^2y}{dt^2} + 4\frac{dy}{dt} + 5y = 0,$$

and the corresponding system is

$$\frac{dy}{dt} = v$$

$$\frac{dv}{dt} = -5y - 4v.$$

(b) Recall that we can read off the characteristic equation of the second-order equation straight from the equation without having to revert to the corresponding system. We obtain

$$\lambda^2 + 4\lambda + 5 = 0.$$

Therefore, the eigenvalues are $\lambda_1 = -2 + i$ and $\lambda_2 = -2 - i$.
To find the eigenvectors associated to the eigenvalue λ_1, we solve the simultaneous system of equations

$$\begin{cases} v = (-2 + i)y \\ -8y - 6v = (-2 + i)v. \end{cases}$$

From the first equation, we immediately see that the eigenvectors associated to this eigenvalue must satisfy $v = (-2+i)y$. Similarly, the eigenvectors associated to the eigenvalue $\lambda_2 = -2-i$ must satisfy the equation $v = (-2 - i)y$.

(c) Since the eigenvalues are complex with negative real part, the equilibrium point at the origin is a spiral sink, and the system is underdamped.

(d) All solutions tend to the origin spiralling in the clockwise direction with period 2π. Admittedly, it is difficult to see these oscillations in the picture.

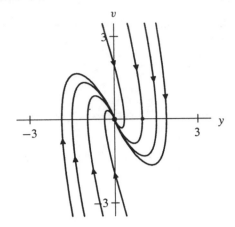

(e) The graph of $y(t)$ initially decreases then oscillates with decreasing amplitude as it tends to 0. Similarly, $v(t)$ initially decreases and becomes negative, then oscillates with decreasing amplitude as it tends to 0.

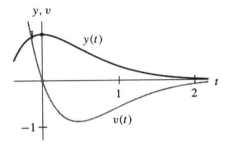

17. (a) The resulting second-order equation is

$$2\frac{d^2y}{dt^2} + 3\frac{dy}{dt} + y = 0,$$

and the corresponding system is

$$\frac{dy}{dt} = v$$
$$\frac{dv}{dt} = -\frac{1}{2}y - \frac{3}{2}v.$$

(b) Recall that we can read off the characteristic equation of the second-order equation straight from the equation without having to revert to the corresponding system. We obtain

$$2\lambda^2 + 3\lambda + 1 = 0.$$

Therefore, the eigenvalues are $\lambda_1 = -1$ and $\lambda_2 = -1/2$.

To find the eigenvectors associated to the eigenvalue λ_1, we solve the simultaneous system of equations

$$\begin{cases} v = -y \\ -\frac{1}{2}y - \frac{3}{2}v = -v. \end{cases}$$

From the first equation, we immediately see that the eigenvectors associated to this eigenvalue must satisfy $v = -y$. Similarly, the eigenvectors associated to the eigenvalue $\lambda_2 = -1/2$ must satisfy the equation $v = -y/2$.

(c) Since the eigenvalues are real and negative, the equilibrium point at the origin is a sink, and the system is overdamped.

(d) We know that all solution curves approach the origin as $t \to \infty$ and, with the exception of those whose initial conditions lie on the line $v = -y$, these solution curves approach the origin tangent to the line $v = -y/2$.

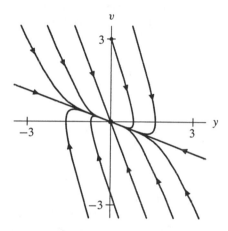

(e) According to the phase plane, $y(t)$ increases initially. Eventually it reaches a maximum value. Then it approaches 0 as $t \to \infty$. Also, $v(t)$ decreases, becomes negative, and then approaches 0 from below. While sketching these graphs, it is useful to remember that $v = dy/dt$.

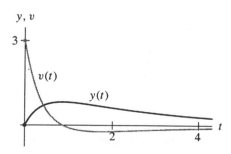

19. **(a)** The resulting second-order equation is

$$2\frac{d^2y}{dt^2} + 3y = 0,$$

and the corresponding system is

$$\frac{dy}{dt} = v$$
$$\frac{dv}{dt} = -\frac{3}{2}y.$$

(b) Recall that we can read off the characteristic equation of the second-order equation straight from the equation without having to revert to the corresponding system. We obtain

$$2\lambda^2 + 3 = 0.$$

Therefore, we have pure imaginary eigenvalues, $\lambda = \pm i\sqrt{3/2}$.
To find the eigenvectors associated to the eigenvalue $\lambda = i\sqrt{3/2}$, we solve the simultaneous system of equations

$$\begin{cases} v = i\sqrt{\frac{3}{2}}\,y \\ -\frac{3}{2}y = i\sqrt{\frac{3}{2}}\,v. \end{cases}$$

From the first equation, we immediately see that the eigenvectors associated to this eigenvalue must satisfy $v = i\sqrt{3/2}\,y$. Similarly, the eigenvectors associated to the eigenvalue $\lambda = -i\sqrt{3/2}$ must satisfy the equation $v = -i\sqrt{3/2}\,y$.

(c) Since the eigenvalues are pure imaginary, the system is undamped. (Of course, we already knew this because $b = 0$.) The natural period is $2\pi/\sqrt{3/2} = 4\pi/\sqrt{6}$.

(d) Since the eigenvalues are pure imaginary, we know that the solution curves are ellipses. At the point $(1, 0)$, $d\mathbf{Y}/dt = (0, -3/2)$. Therefore, we know that the oscillation is clockwise.

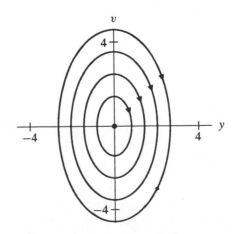

(e)

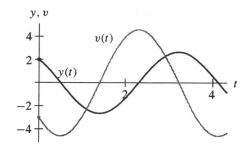

21. **(a)** The second-order equation is

$$\frac{d^2y}{dt^2} + 8\frac{dy}{dt} + 7y = 0,$$

so the characteristic equation is

$$s^2 + 8s + 7 = 0.$$

The roots are $s = -7$ and $s = -1$. The general solution is

$$y(t) = k_1 e^{-7t} + k_2 e^{-t}.$$

(b) To find the particular solution we compute

$$v(t) = -7k_1 e^{-7t} - k_2 e^{-t}.$$

The particular solution satisfies

$$\begin{cases} -1 = y(0) = k_1 + k_2 \\ 5 = v(0) = -7k_1 - k_2. \end{cases}$$

The first equation yields $k_1 = -k_2 - 1$. Substituting into the second we obtain $5 = 6k_2 + 7$, which implies $k_2 = -1/3$. The first equation then yields $k_1 = -2/3$. The particular solution is

$$y(t) = -\tfrac{2}{3}e^{-7t} - \tfrac{1}{3}e^{-t}.$$

(c) The $y(t)$- and $v(t)$-graphs are displayed in the solution of Exercise 13.

23. **(a)** The second-order equation is

$$\frac{d^2y}{dt^2} + 4\frac{dy}{dt} + 5y = 0,$$

so the characteristic equation is

$$s^2 + 4s + 5 = 0.$$

The roots are $s = -2 + i$ and $s = -2 - i$. A complex-valued solution is

$$y_c(t) = e^{(-2+i)t} = e^{-2t}\cos t + i e^{-2t}\sin t.$$

Therefore the general solution is

$$y(t) = k_1 e^{-2t}\cos t + k_2 e^{-2t}\sin t.$$

(b) To find the particular solution we compute

$$v(t) = (-2k_1 + k_2)e^{-2t}\cos t + (-k_1 - 2k_2)e^{-2t}\sin t.$$

The particular solution satisfies

$$\begin{cases} 1 = y(0) = k_1 \\ 0 = v(0) = -2k_1 + k_2. \end{cases}$$

The first equation yields $k_1 = 1$. Substituting into the second we obtain $k_2 = 2$. The particular solution is

$$y(t) = e^{-2t}\cos t + 2e^{-2t}\sin t.$$

(c) The $y(t)$- and $v(t)$-graphs are displayed in the solution of Exercise 15.

25. (a) The second-order equation is

$$2\frac{d^2y}{dt^2} + 3\frac{dy}{dt} + y = 0,$$

so the characteristic equation is

$$2s^2 + 3s + 1 = 0.$$

The roots are $s = -1$ and $s = -1/2$. So the general solution is

$$y(t) = k_1 e^{-t} + k_2 e^{-t/2}.$$

(b) To find the particular solution we compute

$$v(t) = -k_1 e^{-t} - \frac{k_2}{2}e^{-t/2}.$$

The particular solution satisfies

$$\begin{cases} 0 = y(0) = k_1 + k_2 \\ 3 = v(0) = -k_1 - \frac{k_2}{2}. \end{cases}$$

The first equation yields $k_1 = -k_2$. Substituting into the second we obtain $3 = k_2 - k_2/2$, which implies that $k_2 = 6$. The first equation then yields $k_1 = -6$. The particular solution is

$$y(t) = -6e^{-t} + 6e^{-t/2}.$$

(c) The $y(t)$- and $v(t)$-graphs are displayed in the solution of Exercise 17.

27. (a) The second-order equation is

$$2\frac{d^2y}{dt^2} + 3y = 0,$$

so the characteristic equation is

$$2s^2 + 3 = 0.$$

The roots are $s = \sqrt{3/2}\,i$ and $s = -\sqrt{3/2}\,i$. A complex-valued solution is

$$y_c(t) = e^{(\sqrt{3/2}\,i)t} = \cos(\sqrt{3/2}\,t) + i\sin(\sqrt{3/2}\,t).$$

Therefore the general solution is

$$y(t) = k_1\cos(\sqrt{3/2}\,t) + k_2\sin(\sqrt{3/2}\,t).$$

(b) To find the particular solution we compute

$$v(t) = -\sqrt{3/2}\,k_1 \sin(\sqrt{3/2}\,t) + \sqrt{3/2}\,k_2 \cos(\sqrt{3/2}\,t).$$

The particular solution satisfies

$$\begin{cases} 2 = y(0) = k_1 \\ -3 = v(0) = \sqrt{3/2}\,k_2. \end{cases}$$

The first equation yields $k_1 = 2$. Substituting into the second we find $k_2 = -\sqrt{6}$. So the particular solution is

$$y(t) = 2\cos(\sqrt{3/2}\,t) - \sqrt{6}\sin(\sqrt{3/2}\,t).$$

(c) The $y(t)$- and $v(t)$-graphs are displayed in the solution of Exercise 19.

29. Note: We assume that m, k and b are nonnegative—the physically relevant case. All references to graphs and phase portraits are from Section 3.6.

Table 3.1
Possible harmonic oscillators.

name	eigenvalues	parameters	decay rate	phase portrait and graphs
undamped	pure imaginary	$b = 0$	no decay	Figure 3.41
underdamped	complex with negative real part	$b^2 - 4mk < 0$	$e^{-bt/(2m)}$	Figure 3.42
critically damped	only one eigenvalue	$b^2 - 4mk = 0$	$e^{-bt/(2m)}$	Figure 3.34
overdamped	two negative real	$b^2 - 4mk > 0$	$e^{\lambda t}$ where $\lambda = \dfrac{-b + \sqrt{b^2 - 4mk}}{2m}$	Figures 3.43–3.45 and Exercise 13

31. Note that

$$\frac{dy}{dt} = \frac{d}{dt}(y_{re} + i y_{im}) = \frac{dy_{re}}{dt} + i\frac{dy_{im}}{dt}$$

and

$$\frac{d^2y}{dt^2} = \frac{d^2}{dt^2}(y_{re} + i y_{im}) = \frac{d^2 y_{re}}{dt^2} + i\frac{d^2 y_{im}}{dt^2}.$$

Then note that

$$\frac{d^2y}{dt^2} + p\frac{dy}{dt} + qy = \left(\frac{d^2 y_{re}}{dt^2} + p\frac{dy_{re}}{dt} + q y_{re}\right) + i\left(\frac{d^2 y_{im}}{dt^2} + p\frac{dy_{im}}{dt} + q y_{im}\right).$$

Both

$$\frac{d^2 y_{re}}{dt^2} + p\frac{dy_{re}}{dt} + q y_{re} = 0 \quad \text{and} \quad \frac{d^2 y_{im}}{dt^2} + p\frac{dy_{im}}{dt} + q y_{im} = 0$$

because a complex number is zero only if both its real and imaginary parts vanish. In other words, $y_{re}(t)$ and $y_{im}(t)$ are solutions of the original equation.

33. **(a)** If we let $v = dy/dt$, then the corresponding first-order system is

$$\frac{dy}{dt} = v$$

$$\frac{dv}{dt} = -qy - pv,$$

and the corresponding matrix is

$$\mathbf{A} = \begin{pmatrix} 0 & 1 \\ -q & -p \end{pmatrix}.$$

If λ_0 is a repeated eigenvalue, then the characteristic polynomial is

$$\lambda^2 + p\lambda + q = (\lambda - \lambda_0^2) = \lambda^2 - 2\lambda_0\lambda + \lambda_0^2.$$

Consequently, $p = -2\lambda_0$, $q = \lambda_0^2$, and

$$\mathbf{A} = \begin{pmatrix} 0 & 1 \\ -\lambda_0^2 & 2\lambda_0 \end{pmatrix}.$$

(b) To compute the general solution of the corresponding first-order system, we consider an arbitrary initial condition $\mathbf{V}_0 = (y_0, v_0)$ and calculate

$$(\mathbf{A} - \lambda_0\mathbf{I})\mathbf{V}_0 = \begin{pmatrix} -\lambda_0 & 1 \\ -\lambda_0^2 & \lambda_0 \end{pmatrix} \begin{pmatrix} y_0 \\ v_0 \end{pmatrix}$$

$$= \begin{pmatrix} -\lambda_0 y_0 + v_0 \\ -\lambda_0^2 y_0 + \lambda_0 v_0 \end{pmatrix}.$$

The general solution of the first-order system is

$$\mathbf{Y}(t) = e^{\lambda_0 t} \begin{pmatrix} y_0 \\ v_0 \end{pmatrix} + t e^{\lambda_0 t} \begin{pmatrix} -\lambda_0 y_0 + v_0 \\ -\lambda_0^2 y_0 + \lambda_0 v_0 \end{pmatrix}.$$

(c) From the first component of the result in part (b), we obtain the general solution of the original second-order equation in the form

$$y(t) = y_0 e^{\lambda_0 t} + (-\lambda_0 y_0 + v_0) t e^{\lambda_0 t}.$$

(d) Let $k_1 = y_0$ and $k_2 = -\lambda_0 y_0 + v_0$. Clearly, all k_1 are possible. Moreover, once the value of k_1 is determined, k_2 can be determined from v_0 using $k_2 = -\lambda_0 k_1 + v_0$, and v_0 can be determined by k_2 using $v_0 = k_2 + \lambda_0 k_1$. Hence, k_1 and k_2 are arbitrary constants because y_0 and v_0 are arbitrary.

35. The characteristic equation for this harmonic oscillator is

$$s^2 + bs + 3 = 0,$$

and the roots are

$$s_1 = \frac{-b - \sqrt{b^2 - 12}}{2} \quad \text{and} \quad s_2 = \frac{-b + \sqrt{b^2 - 12}}{2}.$$

If $b^2 < 12$, these roots are complex. In this case, all solutions include a factor of the form $e^{(-b/2)t}$, and they tend to the equilibrium at this rate.

If $b^2 > 12$, the roots are real, and the general solution is

$$y(t) = k_1 e^{s_1 t} + k_2 e^{s_2 t}.$$

The first exponential in this expression tends to 0 most quickly, so if $k_2 = 0$, we have solutions that tend to 0 at the rate of $e^{s_1 t}$. This rate is the quickest approach to 0.

The roots are repeated if $b^2 - 12 = 0$, that is, if $b = 2\sqrt{3}$. The fastest approach is then given by a term of the form $e^{-\sqrt{3}t}$.

37. **(a)** Since the fluid causes the object to accelerate as it moves and the force causing this acceleration is proportional to the velocity, the force equation for this "mass-spring" system is

$$m\frac{d^2y}{dt^2} = -ky + b_{mf}\frac{dy}{dt},$$

which can be written as

$$m\frac{d^2y}{dt^2} - b_{mf}\frac{dy}{dt} + ky = 0.$$

(b) The equivalent first-order system is

$$\frac{dy}{dt} = v$$

$$\frac{dv}{dt} = -\frac{k}{m}y + \frac{b_{mf}}{m}v.$$

(c) The characteristic equation is

$$m\lambda^2 - b_{mf}\lambda + k = 0,$$

and the eigenvalues are

$$\frac{b_{mf} \pm \sqrt{b_{mf}^2 - 4mk}}{2m}.$$

Since m, b_{mf}, and k are all positive parameters, the eigenvalues are either positive real numbers or complex numbers with a positive real part. If both eigenvalues are real, then the origin is called an "overstimulated" source. The magnitudes of $y(t)$ and $v(t)$ tend to infinity without oscillation. If the eigenvalues are complex, then the origin is a spiral source and the oscillator is called understimulated. The solutions spiral away from the origin with natural period $4m\pi/\sqrt{b_{mf}^2 - 4mk}$.

39. The differential equation is

$$m\frac{d^2y}{dt^2} + 2y = 0,$$

and the characteristic equation is

$$m\lambda^2 + 2 = 0.$$

Hence, the eigenvalues are $\lambda = \pm i\sqrt{2/m}$. The natural period is $2\pi/\sqrt{2/m} = \pi\sqrt{2m}$. For natural period to be 1, we must have $m = 1/(2\pi^2)$.

EXERCISES FOR SECTION 3.7

1.

Table 3.2
Possibilities for linear systems

type	condition on λ	examples
sink	$\lambda_1 < \lambda_2 < 0$	Sec. 3.7, Fig. 3.52
saddle	$\lambda_1 < 0 < \lambda_2$	Sec. 3.3, Fig. 3.12–3.14
source	$0 < \lambda_1 < \lambda_2$	Sec. 3.3, Fig. 3.19
spiral sink	$\lambda = \alpha \pm i\beta, \alpha < 0, \beta \neq 0$	Sec. 3.1, Fig. 3.2 and 3.4
spiral source	$\lambda = \alpha \pm i\beta, \alpha > 0, \beta \neq 0$	Sec. 3.4, Fig. 3.29–3.30
center	$\lambda_1 = \pm i\beta, \beta \neq 0$	Sec. 3.1, Fig. 3.1 and 3.3
		Sec. 3.4, Fig. 3.28
sink	$\lambda_1 = \lambda_2 < 0$	Sec. 3.5, Fig. 3.35–3.36
(special case)	One line of eigenvectors	
source	$0 < \lambda_1 = \lambda_2$	Sec. 3.5, Ex. 2
(special case)	One line of eigenvectors	
sink	$\lambda_1 = \lambda_2 < 0$	Sec. 3.5, Ex. 23
(special case)	Every vector is eigenvector	
source	$0 < \lambda_1 = \lambda_2$	Sec. 3.5, Ex. 23
(special case)	Every vector is eigenvector	
no name	$\lambda_1 < \lambda_2 = 0$	Sec. 3.5, Fig. 3.39–3.40
no name	$0 = \lambda_1 < \lambda_2$	Sec. 3.5, Ex. 19
no name	$\lambda_1 = \lambda_2 = 0$	Sec. 3.5, Ex. 21
	One line of eigenvectors	
no name	$\lambda_1 = \lambda_2 = 0$	entire plane of equilibrium points
	Every vector is an eigenvector	

3. (a)

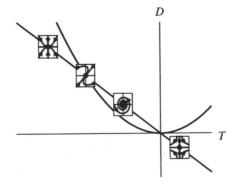

(b) The trace T is $2a$, and the determinant D is $-a$. Therefore, the curve in the trace-determinant plane is $D = -T/2$. This line crosses the parabola $T^2 - 4D = 0$ at two points—at $(T, D) = (0, 0)$ if $a = 0$ and at $(T, D) = (-2, 1)$ if $a = -1$.

The portion of the line for which $a < -1$ corresponds to a positive determinant and a negative trace such that $T^2 - 4D < 0$. The corresponding phase portraits are real sinks. If $a = -1$, we have a sink with repeated eigenvalues. If $-1 < a < 0$, we have complex eigenvalues with negative real parts. Therefore, the phase portraits are spiral sinks. If $a = 0$, we have a degenerate case with an entire line of equilibrium points. Finally, if $a > 0$, the corresponding portion of the line is below the T-axis, and the phase portraits are saddles.

(c) Bifurcations occur at $a = -1$, where we have a sink with repeated eigenvalues, and at $a = 0$, where we have zero as a repeated eigenvalue. For $a = 0$, the y-axis is entirely composed of equilibrium points.

5. (a)

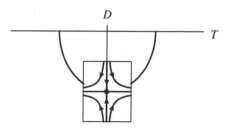

(b) The curve in the trace-determinant plane is the portion of the unit circle centered at 0 that lies in the half-plane $y \le 0$.
A glance at the trace-determinant plane shows that for $-1 < a < 1$, we have a saddle. If $a = 1$, the eigenvalues are 0 and 1. If $a = -1$, the eigenvalues are 0 and -1.

(c) Bifurcations occur only at $a = \pm 1$. For these two special values of a, we have a line of equilibrium points. The nonzero equilibrium points disappear if $-1 < a < 1$.

7. (a)

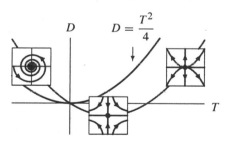

(b) The trace T is $2a$, and the determinant D is $a^2 - a$. Therefore, the curve in the trace-determinant plane is

$$D = a^2 - a$$

$$= \left(\frac{T}{2}\right)^2 - \frac{T}{2}$$

$$= \frac{T^2}{4} - \frac{T}{2}.$$

This curve is a parabola. It meets the repeated-eigenvalue parabola (the parabola $D = T^2/4$) if

$$\frac{T^2}{4} - \frac{T}{2} = \frac{T^2}{4}.$$

Solving this equation yields $T = 0$, which corresponds to $a = 0$.
This curve also meets the T-axis (the line $D = 0$) if

$$\frac{T^2}{4} - \frac{T}{2} = 0,$$

so if $T = 0$ or $T = 2$, then $D = 0$.

From the location of the parabola $D = T^2/4 - T/2$ in the trace-determinant plane, we see that the phase portrait is a spiral sink if $a < 0$ since $T < 0$, a saddle if $0 < a < 1$ since $0 < T < 2$, and a source with distinct real eigenvalues if $a > 1$ since $T > 2$.

(c) Bifurcations occur at $a = 0$, where we have repeated zero eigenvalues, and at $a = 1$, where we have a single zero eigenvalue.

9. The eigenvalues are roots of the equation $\lambda^2 - 2a\lambda + a^2 - b^2 = 0$. These roots are

$$a \pm \sqrt{b^2} = a \pm |b|.$$

So we have a repeated zero eigenvalue if $a = b = 0$.

If $a = \pm b$, then one of the eigenvalues is 0, and as long as $a \neq 0$ (so $b \neq 0$), the other eigenvalue is nonzero.

The eigenvalues are repeated (both equal to a) if $b = 0$. The eigenvalues are never complex since $\sqrt{b^2} \geq 0$.

If $a > |b|$, then $a \pm |b| > 0$, so we have a source with real eigenvalues. If $a < 0$ and $-a > |b|$, then $a \pm |b| < 0$, so we have a sink with real eigenvalues. In all other cases we have a saddle.

11. (a) This second-order equation is equivalent to the system

$$\frac{dy}{dt} = v$$
$$\frac{dv}{dt} = -3y - bv.$$

Therefore, $T = -b$ and $D = 3$. So the corresponding curve in the trace-determinant plane is $D = 3$.

(b)

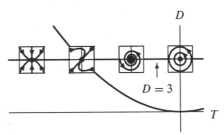

(c) The line $D = 3$ in the trace-determinant plane crosses the repeated-eigenvalue parabola $D = T^2/4$ if $b^2 = 12$, which implies that $b = 2\sqrt{3}$ since b is a nonnegative parameter. If $b = 0$, we have pure imaginary eigenvalues—the undamped case. If $0 < b < 2\sqrt{3}$, the eigenvalues are complex with a negative real part—the underdamped case. If $b = 2\sqrt{3}$, the eigenvalues are repeated and negative—the critically damped case. Finally, if $b > 2\sqrt{3}$, the eigenvalues are real and negative—the overdamped case.

13. **(a)** The second-order equation reduces to the first-order system

$$\frac{dy}{dt} = v$$

$$\frac{dv}{dt} = -\frac{2}{m}y + \frac{1}{m}v.$$

Hence $T = -1/m$, $D = 2/m$, and as the parameter m varies, the systems move along the line $D = -2T$ in the second quadrant of the trace-determinant plane.

(b)

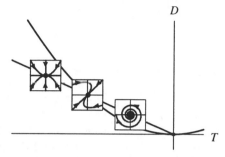

(c) The line $D = -2T$ intersects the repeated-eigenvalue parabola $D = T^2/4$ at the point (T, D) that satisfies $-2T = T^2/4$. We have

$$\frac{T^2}{4} + 2T = T\left(\frac{T}{4} + 2\right) = 0,$$

which yields $T = 0$ or $T = -8$.

For $-8 < T < 0$, the system is underdamped; for $T = -8$, the system is critically damped; and for $T < -8$, the system is overdamped. Since $T = -1/m$, the system is overdamped if $0 < m < 1/8$; it is critically damped if $m = 1/8$; and it is underdamped if $m > 1/8$.

EXERCISES FOR SECTION 3.8

1. To check that the given vector-valued functions are solutions, we differentiate each coordinate and check that the system is satisfied. More precisely, we must check that $dx/dt = 0.1y$, $dy/dt = 0.2z$, and $dz/dt = 0.4x$.

To check that these equations are satisfied, we simply differentiate and simplify. For example, to check that $\mathbf{Y}_2(t)$ is a solution, we first differentiate the x-coordinate, and we obtain

$$\frac{dx}{dt} = -0.1\,e^{-0.1t}\left(-\cos\left(\sqrt{0.03}\,t\right) - \sqrt{3}\,\sin\left(\sqrt{0.03}\,t\right)\right) +$$

$$e^{-0.1t}\left(\sqrt{0.03}\,\sin\left(\sqrt{0.03}\,t\right) - \sqrt{3}\,\sqrt{0.03}\,\cos\left(\sqrt{0.03}\,t\right)\right)$$

$$= e^{-0.1t}\left(\left(0.1 - \sqrt{0.09}\right)\cos\left(\sqrt{0.03}\,t\right) + \left(0.1\sqrt{3} + \sqrt{0.03}\right)\sin\left(\sqrt{0.03}\,t\right)\right)$$

$$= e^{-0.1t}\left(-0.2\cos\left(\sqrt{0.03}\,t\right) + 0.2\sqrt{3}\,\sin\left(\sqrt{0.03}\,t\right)\right)$$

$$= 0.1\,e^{-0.1t}\left(-2\cos\left(\sqrt{0.03}\,t\right) + 2\sqrt{3}\,\sin\left(\sqrt{0.03}\,t\right)\right).$$

Note that this last expression is just $0.1 y(t)$. Hence, the first component of the differential equation is satisfied.

In order to complete the verification that $\mathbf{Y}_2(t)$ is a solution, we must also verify the equations for dy/dt and dz/dt. The calculations are similar.

3. **(a)** Suppose

$$k_1 \begin{pmatrix} 1 \\ 2 \\ 1 \end{pmatrix} + k_2 \begin{pmatrix} 1 \\ 3 \\ 1 \end{pmatrix} + k_3 \begin{pmatrix} 1 \\ 4 \\ 1 \end{pmatrix} = \begin{pmatrix} 0 \\ 0 \\ 0 \end{pmatrix}.$$

We obtain the simultaneous equations

$$\begin{cases} k_1 + k_2 + k_3 = 0 \\ 2k_1 + 3k_2 + 4k_3 = 0. \end{cases}$$

We have two equations with three unknowns. Therefore, we cannot uniquely determine the values of k_1, k_2, and k_3. In other words, we can find infinitely many triples (k_1, k_2, k_3) such that $k_1 \mathbf{Y}_1 + k_2 \mathbf{Y}_2 + k_3 \mathbf{Y}_3 = 0$. For example, $k_1 = -1$, $k_2 = 2$, and $k_3 = -1$ is one such triple.

(b) Suppose

$$k_1 \begin{pmatrix} 2 \\ 0 \\ 1 \end{pmatrix} + k_2 \begin{pmatrix} 3 \\ 2 \\ 2 \end{pmatrix} + k_3 \begin{pmatrix} 1 \\ -2 \\ -3 \end{pmatrix} = \begin{pmatrix} 0 \\ 0 \\ 0 \end{pmatrix}.$$

In matrix notation, we can write this as

$$\mathbf{A} \begin{pmatrix} k_1 \\ k_2 \\ k_3 \end{pmatrix} = \begin{pmatrix} -2 & 3 & 0 \\ 3 & -2 & 0 \\ 0 & 0 & -1 \end{pmatrix} \begin{pmatrix} k_1 \\ k_2 \\ k_3 \end{pmatrix} = \begin{pmatrix} 0 \\ 0 \\ 0 \end{pmatrix}.$$

Since $\det \mathbf{A} = -12 \neq 0$, $\mathbf{A}$ has an inverse matrix, and by multiplying the inverse matrix, we obtain $(k_1, k_2, k_3) = (0, 0, 0)$. Hence, the vectors are linearly independent.

(c) Suppose

$$k_1 \begin{pmatrix} 1 \\ 2 \\ 0 \end{pmatrix} + k_2 \begin{pmatrix} 0 \\ 1 \\ 2 \end{pmatrix} + k_3 \begin{pmatrix} 2 \\ 0 \\ 1 \end{pmatrix} = \begin{pmatrix} 0 \\ 0 \\ 0 \end{pmatrix}.$$

In scalar form, this vector equation is equivalent to the scalar equations

$$k_1 + k_3 = 0$$
$$2k_1 + k_2 = 0$$
$$2k_2 + k_3 = 0.$$

From the third equation, $-2k_3 = k_1$, and substitution of this equation into the first equation yields $k_1 = 0$. Then, $k_3 = 0$, and using the second equation, we conclude that $k_2 = 0$. Therefore, the three vectors are independent.

(d) Suppose

$$k_1 \begin{pmatrix} -3 \\ \pi \\ 1 \end{pmatrix} + k_2 \begin{pmatrix} 0 \\ 1 \\ 0 \end{pmatrix} + k_3 \begin{pmatrix} -2 \\ -2 \\ -2 \end{pmatrix} = \begin{pmatrix} 0 \\ 0 \\ 0 \end{pmatrix}.$$

This vector equation can be written as the simultaneous equations

$$-3k_1 - 2k_3 = 0$$
$$\pi k_1 + k_2 - 2k_3 = 0$$
$$k_1 - 2k_3 = 0.$$

From the third equation, $2k_3 = k_1$ and substitution of this equation into the first equation yields $k_1 = 0$. Then, $k_3 = 0$, and using the second equation, we obtain $k_2 = 0$. Therefore, the three vectors are independent.

5. (a) The characteristic equation is

$$\det(\mathbf{A} - \lambda \mathbf{I}) = (-2 - \lambda)(-2 - \lambda)(-1 - \lambda) - (3)(3)(-1 - \lambda) = 0,$$

which reduces to

$$-(\lambda + 1)(\lambda + 5)(\lambda - 1) = 0.$$

Therefore, the eigenvalues are $\lambda = \pm 1$ and $\lambda = -5$.

(b) Writing the differential equation in coordinates, one obtains

$$\frac{dx}{dt} = -2x + 3y$$
$$\frac{dy}{dt} = 3x - 2y$$
$$\frac{dz}{dt} = -z.$$

Since dx/dt and dy/dt do not depend on z and dz/dt does not depend on x or y, the system decouples into a two-dimensional system in the xy-plane and a one-dimensional system on the z-axis.

(c) In the xy-plane, the characteristic equation is $(-2 - \lambda)^2 - 9 = \lambda^2 + 4\lambda - 5 = 0$, and the eigenvalues are $\lambda = -5$ and $\lambda = 1$. Therefore, the system is a saddle in the xy-plane. The eigenvectors (x, y, z) for $\lambda = -5$ satisfy the equations $y = -x$ and $z = 0$, and the eigenvectors for $\lambda = 1$ satisfy the equations $x = y$ and $z = 0$.

The z-axis is the phase line for the equation $dz/dt = -z$. Therefore, there is a single equilibrium point at the origin, and every solution curve lying on the z-axis is asymptotic to 0 as $t \to \infty$.

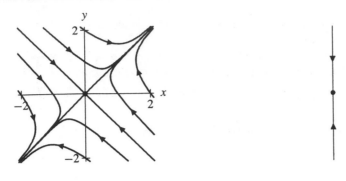

xy-phase plane   z-phase line

(d) The xy-phase plane and the z-phase line can be combined to obtain the xyz-phase space.

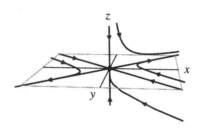

7. **(a)** The characteristic equation is

$$\det(\mathbf{A} - \lambda \mathbf{I}) = (1 - \lambda)((2 - \lambda)(2 - \lambda) - 1) = 0,$$

which simplifies to

$$-(\lambda - 3)(\lambda - 1)(\lambda - 1) = 0.$$

Therefore, the eigenvalues are $\lambda = 1$ and $\lambda = 3$.

(b) Writing the differential equation in coordinates, one obtains

$$\frac{dx}{dt} = x$$

$$\frac{dy}{dt} = 2y - z$$

$$\frac{dz}{dt} = -y + 2z.$$

Since dy/dt and dz/dt do not depend on x, and dx/dt does not depend on y or z, the system decouples into a two-dimensional system in the yz-plane and a one-dimensional system on the x-axis.

(c) In yz-plane, the characteristic equation is $(2-\lambda)^2 - 1 = (\lambda - 3)(\lambda - 1) = 0$, and the eigenvalues are $\lambda = 1$ and $\lambda = 3$. Therefore, in the yz-phase plane, the system is a source. The eigenvectors

(x, y, z) for $\lambda = 1$ satisfy the equations $x = 0$ and $y = z$, and the eigenvectors for $\lambda = 3$ satisfy the equations $x = 0$ and $y = -z$.

The x-axis is the phase line for the equation $dx/dt = x$. Therefore, there is a single equilibrium point at the origin, and every solution curve lying on the x-axis tends to infinity as $t \to \infty$.

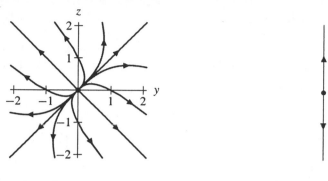

yz-phase plane x-phase line

(d) The x-phase line and the yz-phase plane can be combined to obtain the xyz-phase space.

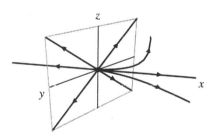

9. (Note: This exercise can also be done by taking the complex conjugate of both sides of the equation $p(a + ib) = 0$.)

Using the definition of $p(x)$, we compute

$$p(a + ib) = (\alpha a^3 - 3\alpha ab^2 + \beta a^2 - \beta b^2 + \gamma a + \delta) + i(3\alpha a^2 b - \alpha b^3 + 2\beta ab + \gamma b)$$

and

$$p(a - bi) = (\alpha a^3 - 3\alpha ab^2 + \beta a^2 - \beta b^2 + \gamma a + \delta) - i(3\alpha a^2 b - \alpha b^3 + 2\beta ab + \gamma b).$$

Since $p(a + bi) = 0$,

$$\alpha a^3 - 3\alpha ab^2 + \beta a^2 - \beta b^2 + \gamma a + \delta = 3\alpha a^2 b - \alpha b^3 + 2\beta ab + \gamma b = 0.$$

Therefore, $p(a - ib) = 0$, and $a - ib$ is also a root.

11. **(a)** The characteristic equation is

$$(-2 - \lambda)(-2 - \lambda)(1 - \lambda) = 0,$$

and the eigenvalues are $\lambda = -2$ and $\lambda = 1$.

(b) In the xy-plane, the system has a repeated eigenvalue, $\lambda = -2$, and the eigenvectors (x, y, z) associated to that eigenvalue satisfy the equations $y = z = 0$. The origin is a sink in the xy-phase plane.

The z-axis is the phase line for the equation $dz/dt = z$. Therefore, there is a single equilibrium point at the origin, and every solution curve lying on the z-axis tends to infinity as $t \to \infty$.

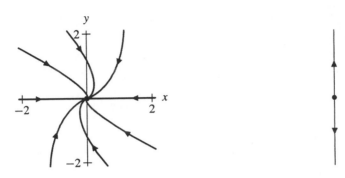

xy-phase plane   z-phase line

(c) Combining the xy-phase plane and z-phase line, we obtain a picture of the xyz-phase space.

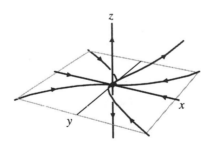

13. **(a)** The characteristic equation is

$$(-1 - \lambda)(-4 - \lambda)(-\lambda) - 4(-\lambda) = -\lambda^2(\lambda + 5) = 0.$$

One eigenvalue is $\lambda = -5$, and the other eigenvalue, $\lambda = 0$, is a repeated eigenvalue.

(b) In xy-plane, the eigenvalues are $\lambda = 0$ and $\lambda = -5$. The eigenvectors (x, y, z) associated to the eigenvalue $\lambda = 0$ satisfy the equations $z = 0$ and $-x + 2y = 0$. The eigenvectors associated to the eigenvalue $\lambda = -5$ satisfy the equations $z = 0$ and $4x + 2y = 0$. Hence, the line $y = x/2$ is a line of equilibrium points, and every solution curve lying in the xy-plane tends toward one of the equilibrium points on $y = x/2$ as $t \to \infty$.

The z-axis is the phase line for $dz/dt = 0$. Note that this line consists entirely of equilibrium points.

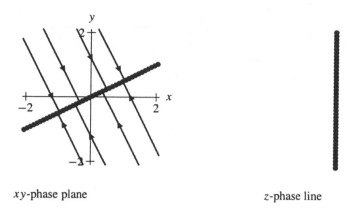

xy-phase plane   z-phase line

(c) Combining the xy-phase plane and z-phase line, we obtain a picture of the xyz-phase space.

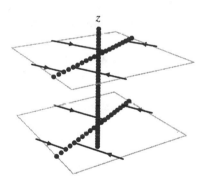

15. **(a)** The characteristic equation is $-\lambda^3 = 0$. Consequently, there is only one eigenvalue, $\lambda = 0$.

(b) For $\lambda = 0$, the eigenvectors (x, y, z) must satisfy both $y = 0$ and $z = 0$. Therefore, the x-axis is both a line of eigenvectors and a line of equilibrium points.

(c) Since $dz/dt = 0$, $z(t)$ is a constant function. That is, if $z(0) = z_0$, then $z(t) = z_0$ for all t. Since $dy/dt = z$ and z is constant, we have $y(t) = z_0 t + y_0$, where $y(0) = y_0$ is the initial condition for $y(t)$. Finally, since $dx/dt = y = z_0 t + y_0$, $x(t) = z_0 t^2/2 + y_0 t + x_0$, where $x(0) = x_0$ is the initial condition for $x(t)$.

For $z_0 = 0$, $y(t)$ is constant, and the solution curves lie on straight lines parallel to the x-axis. For $y_0 > 0$, $x(t)$ is increasing, and for $y_0 < 0$, $x(t)$ is decreasing. For $z_0 \neq 0$, $x(t)$ is quadratic in y. Therefore, solution curves that satisfy the initial condition $z(0) = z_0$ stay on the plane $z = z_0$ and lie on a parabola.

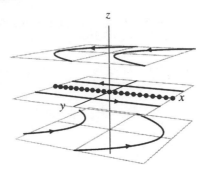

17. **(a)** We compute

$$\mathbf{AV}_1 = \mathbf{A}\begin{pmatrix} 1 \\ 1 \\ 0 \end{pmatrix} = \begin{pmatrix} -1 \\ -1 \\ 0 \end{pmatrix} = -1\begin{pmatrix} 1 \\ 1 \\ 0 \end{pmatrix}.$$

Hence, $\mathbf{V}_1$ is an eigenvector associated to the eigenvalue -1.

(b) The characteristic equation is

$$(-4 - \lambda)((-1 - \lambda)(-\lambda) + 5) + 15 = (\lambda + 1)(\lambda^2 + 4\lambda + 5) = 0.$$

The eigenvalues are $\lambda = -1$ and $\lambda = -2 \pm i$.

(c) Since one eigenvalue is a negative real number and the other two are complex with a negative real part, the system is a spiral sink.

(d) To determine all eigenvectors $\mathbf{V}_2$ associated to the eigenvalue $\lambda = -2 + i$, we must solve the vector equation $\mathbf{AV}_2 = (-2+i)\mathbf{V}_2$. This vector equation is equivalent to the three simultaneous equations

$$\begin{cases} -4x + 3y = (-2+i)x \\ -y + z = (-2+i)y \\ -x + 3y - z = (-2+i)z. \end{cases}$$

Then, all complex eigenvectors $\mathbf{V}_2 = (x, y, z)$ must satisfy the equations $y = (2+i)x/3$ and $z = (-3+i)x/3$. One such eigenvector is $(3, 2+i, -3+i) = (3, 2, -3) + i(0, 1, 1)$. Taking real and imaginary parts, we obtain two vectors, $(3, 2, -3)$ and $(0, 1, 1)$, that determine a plane on which the solutions spiral toward the origin with natural period 2π.

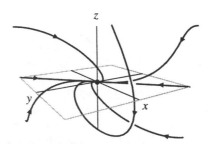

19. **(a)** If Glen makes a profit, z is positive. Since the coefficients of z in the equations for dx/dt and dy/dt are positive, these terms contribute positively to dx/dt and dy/dt. In other words, Glen's profitability helps Paul and Bob be profitable (and they need all the help they can get).

(b) Since dz/dt does not have either an x or a y term, the values of x and y do not contribute to dz/dt. Hence, the profitability of either Paul or Bob makes no difference to Glen's future profits.

21. **(a)** For $z = 0$, $dz/dt = 0$ and, therefore, $z(t) = 0$ for all t. Consequently, we can analyze this system as if it has only two dependent variables, x and y. We have

$$\frac{dx}{dt} = -y$$
$$\frac{dy}{dt} = -x.$$

This system is a saddle in the xy-plane.

The eigenvalues for the xy-system are ± 1, and the eigenvectors for the eigenvalue -1 satisfy the equation $x = y$. Since we are assuming $x(0) = y(0)$, the given solution tends to the origin along the line $x = y$ in the xy-plane.

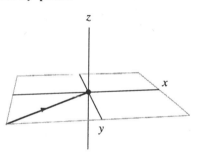

(b) Since $x(0) = y(0)$ and $z(0) = 0$ is an initial condition that is an eigenvector associated to the eigenvalue -1, we know that $x(t) = y(t)$ and $z(t) = 0$ for all t. We also know that both $x(t)$ and $y(t)$ decay to 0 like the function e^{-t}.

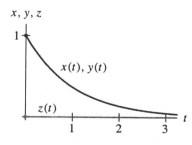

The graphs of $x(t)$ and $y(t)$ are identical, and $z(t) = 0$ for all t.

(c) Glen continues to break even. Both Paul's and Bob's profits tend to the break-even point as $t \to \infty$.

CHAPTER 4

Forcing and Resonance

EXERCISES FOR SECTION 4.1

1. To compute the general solution of the unforced equation, we use the method of Section 3.6. The characteristic polynomial is

$$s^2 + 6s + 8,$$

so the eigenvalues are $s = -2$ and $s = -4$. Hence, the general solution of the homogeneous equation is

$$k_1 e^{-2t} + k_2 e^{-4t}.$$

To find a particular solution of the forced equation, we guess $y_p(t) = ke^{-t}$. Substituting into the left-hand side of the differential equation gives

$$\frac{d^2 y_p}{dt^2} + 6\frac{dy_p}{dt} + 8y_p = ke^{-t} - 6ke^{-t} + 8ke^{-t}$$
$$= 3ke^{-t}.$$

In order for $y_p(t)$ to be a solution of the forced equation, we must take $k = 1/3$. The general solution of the forced equation is

$$y(t) = k_1 e^{-2t} + k_2 e^{-4t} + \tfrac{1}{3}e^{-t}.$$

3. To compute the general solution of the unforced equation, we use the method of Section 3.6. The characteristic polynomial is

$$s^2 + 7s + 12,$$

so the eigenvalues are $s = -3$ and $s = -4$. Hence, the general solution of the homogeneous equation is

$$k_1 e^{-3t} + k_2 e^{-4t}.$$

To find a particular solution of the forced equation, we guess $y_p(t) = ke^{-2t}$. Substituting into the left-hand side of the differential equation gives

$$\frac{d^2 y_p}{dt^2} + 7\frac{dy_p}{dt} + 12y_p = 4ke^{-2t} - 14ke^{-2t} + 12ke^{-2t}$$
$$= 2ke^{-2t}.$$

In order for $y_p(t)$ to be a solution of the forced equation, we must take $k = 3/2$. The general solution of the forced equation is

$$y(t) = k_1 e^{-3t} + k_2 e^{-4t} + \tfrac{3}{2}e^{-2t}.$$

5. To compute the general solution of the unforced equation, we use the method of Section 3.6. The characteristic polynomial is

$$s^2 + 4s + 13,$$

so the eigenvalues are $s = -2 \pm 3i$. Hence, the general solution of the homogeneous equation is

$$k_1 e^{-2t} \cos 3t + k_2 e^{-2t} \sin 3t.$$

To find a particular solution of the forced equation, we guess $y_p(t) = ke^{-2t}$. Substituting into the left-hand side of the differential equation gives

$$\frac{d^2 y_p}{dt^2} + 4\frac{dy_p}{dt} + 13y_p = 4ke^{-2t} - 8ke^{-2t} + 13ke^{-2t}$$

$$= 9ke^{-2t}.$$

In order for $y_p(t)$ to be a solution of the forced equation, we must take $k = -1/3$. The general solution of the forced equation is

$$y(t) = k_1 e^{-2t} \cos 3t + k_2 e^{-2t} \sin 3t - \tfrac{1}{3} e^{-2t}.$$

7. To compute the general solution of the unforced equation, we use the method of Section 3.6. The characteristic polynomial is

$$s^2 - 2s - 3,$$

so the eigenvalues are $s = -1$ and $s = 3$. Hence, the general solution of the homogeneous equation is

$$k_1 e^{-t} + k_2 e^{3t}.$$

To find a particular solution of the forced equation, a reasonable looking guess is $y_p(t) = ke^{3t}$. However, this guess is a solution of the homogeneous equation, so it is doomed to fail. We make the standard second guess of $y_p(t) = kte^{3t}$. Substituting into the left-hand side of the differential equation gives

$$\frac{d^2 y_p}{dt^2} - 2\frac{dy_p}{dt} - 3y_p = (6ke^{3t} + 9kte^{3t}) - 2(ke^{3t} + 3kte^{3t}) - 3kte^{3t}$$

$$= 4ke^{3t}.$$

In order for $y_p(t)$ to be a solution of the forced equation, we must take $k = 1/4$. The general solution of the forced equation is

$$y(t) = k_1 e^{-t} + k_2 e^{3t} + \tfrac{1}{4}te^{3t}.$$

9. This is the same equation as in Exercise 1. The general solution is

$$y(t) = k_1 e^{-2t} + k_2 e^{-4t} + \tfrac{1}{3}e^{-t}.$$

To find the solution with the initial conditions $y(0) = y'(0) = 0$, we compute

$$y'(t) = -2k_1 e^{-2t} - 4k_2 e^{-4t} - \tfrac{1}{3}e^{-t}.$$

Then we evaluate at $t = 0$ and obtain the simultaneous equations

$$\begin{cases} k_1 + k_2 + \tfrac{1}{3} = 0 \\ -2k_1 - 4k_2 - \tfrac{1}{3} = 0. \end{cases}$$

Solving, we have $k_1 = -1/2$ and $k_2 = 1/6$, so the solution of the initial-value problem is

$$y(t) = -\tfrac{1}{2}e^{-2t} + \tfrac{1}{6}e^{-4t} + \tfrac{1}{3}e^{-t}.$$

11. This is the same equation as Exercise 5. The general solution is

$$y(t) = k_1 e^{-2t} \cos 3t + k_2 e^{-2t} \sin 3t - \tfrac{1}{3} e^{-2t}.$$

To find the solution with the initial conditions $y(0) = y'(0) = 0$, we compute

$$y'(t) = -2k_1 e^{-2t} \cos 3t - 3k_1 e^{-2t} \sin 3t - 2k_2 e^{-2t} \sin 3t + 3k_2 e^{-2t} \cos 3t + \tfrac{2}{3} e^{-2t}.$$

Then we evaluate at $t = 0$ and obtain the simultaneous equations

$$\begin{cases} k_1 - \tfrac{1}{3} = 0 \\ -2k_1 + 3k_2 + \tfrac{2}{3} = 0. \end{cases}$$

Solving, we have $k_1 = 1/3$ and $k_2 = 0$, so the solution of the initial-value problem is

$$y(t) = \tfrac{1}{3} e^{-2t} \cos 3t - \tfrac{1}{3} e^{-2t}.$$

13. **(a)** The characteristic polynomial of the unforced equation is

$$s^2 + 4s + 3.$$

So the eigenvalues are $s = -1$ and $s = -3$, and the general solution of the unforced equation is

$$k_1 e^{-t} + k_2 e^{-3t}.$$

To find a particular solution of the forced equation, we guess $y_p(t) = k e^{-t/2}$. Substituting $y_p(t)$ into the left-hand side of the differential equation gives

$$\frac{d^2 y_p}{dt^2} + 4\frac{dy_p}{dt} + 3y_p = \tfrac{1}{4} k e^{-t/2} - 2k e^{-t/2} + 3k e^{-t/2}$$
$$= \tfrac{5}{4} k e^{-t/2}.$$

So $k = 4/5$ yields a solution of the forced equation.
The general solution of the forced equation is therefore

$$y(t) = k_1 e^{-t} + k_2 e^{-3t} + \tfrac{4}{5} e^{-t/2}.$$

(b) The derivative of the general solution is

$$y'(t) = -k_1 e^{-t} - 3k_2 e^{-3t} - \tfrac{2}{5} e^{-t/2}.$$

To find the solution with $y(0) = y'(0) = 0$, we evaluate at $t = 0$ and obtain the simultaneous equations

$$\begin{cases} k_1 + k_2 + \tfrac{4}{5} = 0 \\ -k_1 - 3k_2 - \tfrac{2}{5} = 0. \end{cases}$$

Solving, we find that $k_1 = -1$ and $k_2 = 1/5$, so the solution of the initial-value problem is

$$y(t) = -e^{-t} + \tfrac{1}{5} e^{-3t} + \tfrac{4}{5} e^{-t/2}.$$

(c) Every solution tends to zero as t increases. Of the three terms that sum to the general solution, $\tfrac{4}{5} e^{-t/2}$ dominates when t is large, so all solutions are approximately $\tfrac{4}{5} e^{-t/2}$ for t large.

15. **(a)** The characteristic polynomial of the unforced equation is

$$s^2 + 4s + 3.$$

So the eigenvalues are $s = -1$ and $s = -3$, and the general solution of the unforced equation is

$$k_1 e^{-t} + k_2 e^{-3t}.$$

To find a particular solution of the forced equation, we guess $y_p(t) = ke^{-4t}$. Substituting $y_p(t)$ into the left-hand side of the differential equation gives

$$\frac{d^2 y_p}{dt^2} + 4\frac{dy_p}{dt} + 3y_p = 16ke^{-4t} - 16ke^{-4t} + 3ke^{-4t}$$

$$= 3ke^{-4t}.$$

So $k = 1/3$ yields a solution of the forced equation.
The general solution of the forced equation is therefore

$$y(t) = k_1 e^{-t} + k_2 e^{-3t} + \tfrac{1}{3}e^{-4t}.$$

(b) The derivative of the general solution is

$$y'(t) = -k_1 e^{-t} - 3k_2 e^{-3t} - \tfrac{4}{3}e^{-4t}.$$

To find the solution with $y(0) = y'(0) = 0$, we evaluate at $t = 0$ and obtain the simultaneous equations

$$\begin{cases} k_1 + k_2 + \tfrac{1}{3} = 0 \\ -k_1 - 3k_2 - \tfrac{4}{3} = 0. \end{cases}$$

Solving, we find that $k_1 = 1/6$ and $k_2 = -1/2$, so the solution of the initial-value problem is

$$y(t) = \tfrac{1}{6}e^{-t} - \tfrac{1}{2}e^{-3t} + \tfrac{1}{3}e^{-4t}.$$

(c) In the general solution, all three terms tend to zero, so the solution tends to zero. We can say a little more by noting that the term $k_1 e^{-t}$ is much larger (provided $k_1 \neq 0$). Hence, most solutions tend to zero at the rate of e^{-t}. If $k_1 = 0$, then solutions tend to zero at the rate of e^{-3t} provided $k_2 \neq 0$.

17. **(a)** The characteristic polynomial of the unforced equation is

$$s^2 + 4s + 20.$$

So the eigenvalues are $s = -2 \pm 4i$, and the general solution of the unforced equation is

$$k_1 e^{-2t} \cos 4t + k_2 e^{-2t} \sin 4t.$$

To find a particular solution of the forced equation, we guess $y_p(t) = ke^{-2t}$. Substituting $y_p(t)$ into the left-hand side of the differential equation gives

$$\frac{d^2 y_p}{dt^2} + 4\frac{dy_p}{dt} + 20y_p = 4ke^{-2t} - 8ke^{-2t} + 20ke^{-2t}$$

$$= 16ke^{-2t}.$$

So $k = 1/16$ yields a solution of the forced equation.
The general solution of the forced equation is therefore

$$y(t) = k_1 e^{-2t} \cos 4t + k_2 e^{-2t} \sin 4t + \tfrac{1}{16} e^{-2t}.$$

(b) The derivative of the general solution is

$$y'(t) = -k_1 e^{-2t} \cos 4t - 4k_1 e^{-2t} \sin 4t$$
$$-2k_2 e^{-2t} \sin 4t + 4k_2 e^{-2t} \cos 4t - \tfrac{1}{8} e^{-2t}.$$

To find the solution with $y(0) = y'(0) = 0$, we evaluate at $t = 0$ and obtain the simultaneous equations

$$\begin{cases} k_1 + \tfrac{1}{16} = 0 \\ -2k_1 + 4k_2 - \tfrac{1}{8} = 0. \end{cases}$$

Solving, we find that $k_1 = -1/16$ and $k_2 = 0$, so the solution of the initial-value problem is

$$y(t) = -\tfrac{1}{16} e^{-2t} \cos 4t + \tfrac{1}{16} e^{-2t}.$$

(c) Every solution tends to zero like e^{-2t} and all but one exponential solution oscillates with frequency $2/\pi$.

19. The natural guesses of $y_p(t) = ke^{-t}$ and $y_p(t) = kte^{-t}$ fail to be solutions of the forced equation because they are both solutions of the unforced equation. (The characteristic polynomial of the unforced equation is

$$s^2 + 2s + 1,$$

which has -1 as a double root.)

So we guess $y_p(t) = kt^2 e^{-t}$. Substituting this guess into the left-hand side of the differential equation gives

$$\frac{d^2 y_p}{dt^2} + 2\frac{dy_p}{dt} + y_p = (2ke^{-t} - 4kte^{-t} + kt^2 e^{-t}) + 2(2kte^{-t} - kt^2 e^{-t}) + kt^2 e^{-t}$$
$$= 2ke^{-t}.$$

So $k = 1/2$ yields the solution

$$y_p(t) = \tfrac{1}{2} t^2 e^{-t}.$$

From the characteristic polynomial, we know that the general solution of the unforced equation is

$$k_1 e^{-t} + k_2 t e^{-t}.$$

Consequently, the general solution of the forced equation is

$$y(t) = k_1 e^{-t} + k_2 t e^{-t} + \tfrac{1}{2} t^2 e^{-t}.$$

21. **(a)** The characteristic polynomial of the unforced equation is

$$s^2 + 6s + 8.$$

So the eigenvalues are $s = -2$ and $s = -4$, and the general solution of the unforced equation is

$$k_1 e^{-2t} + k_2 e^{-4t}.$$

To find one solution of the forced equation, we guess the constant function $y_p(t) = k$. Substituting $y_p(t)$ into the left-hand side of the differential equation, we obtain

$$\frac{d^2 y_p}{dt^2} + 6\frac{dy_p}{dt} + 8y_p = 0 + 6 \cdot 0 + 8k = 8k.$$

Hence, $k = 5/8$ yields a solution of the forced equation. The general solution of the forced equation is

$$y(t) = k_1 e^{-2t} + k_2 e^{-4t} + \tfrac{5}{8}.$$

(b) To find the solution satisfying the initial conditions $y(0) = y'(0) = 0$, we compute the derivative of the general solution

$$y'(t) = -2k_1 e^{-2t} - 4k_2 e^{-4t}.$$

Using the initial conditions and evaluating $y(t)$ and $y'(t)$ at $t = 0$, we obtain the simultaneous equations

$$\begin{cases} k_1 + k_2 + \tfrac{5}{8} = 0 \\ -2k_1 - 4k_2 = 0. \end{cases}$$

Solving for k_1 and k_2 gives $k_1 = -5/4$ and $k_2 = 5/8$. The solution of the initial-value problem is

$$y(t) = -\tfrac{5}{4}e^{-2t} + \tfrac{5}{8}e^{-4t} + \tfrac{5}{8}.$$

23. **(a)** The characteristic polynomial of the unforced equation is

$$s^2 + 2s + 10.$$

So the eigenvalues are $s = -1 \pm 3i$, and the general solution of the unforced equation is

$$k_1 e^{-t} \cos 3t + k_2 e^{-t} \sin 3t.$$

To find one solution of the forced equation, we guess the constant function $y_p(t) = k$. Substituting $y_p(t)$ into the left-hand side of the differential equation, we obtain

$$\frac{d^2 y_p}{dt^2} + 2\frac{dy_p}{dt} + 10y_p = 0 + 2 \cdot 0 + 10k = 10k.$$

Hence, $k = 1$ yields a solution of the forced equation. The general solution of the forced equation is

$$y(t) = k_1 e^{-t} \cos 3t + k_2 e^{-t} \sin 3t + 1.$$

(b) To find the solution satisfying the initial conditions $y(0) = y'(0) = 0$, we compute the derivative of the general solution

$$y'(t) = -k_1 e^{-t} \cos 3t - 3k_1 e^{-t} \sin 3t - k_2 e^{-t} \sin 3t + 3k_2 e^{-t} \cos 3t.$$

Using the initial conditions and evaluating $y(t)$ and $y'(t)$ at $t = 0$, we obtain the simultaneous equations

$$\begin{cases} k_1 + 1 = 0 \\ -k_1 + 3k_2 = 0. \end{cases}$$

Solving for k_1 and k_2 gives $k_1 = -1$ and $k_2 = -1/3$. The solution of the initial-value problem is

$$y(t) = -e^{-t} \cos 3t - \tfrac{1}{3} e^{-t} \sin 3t + 1.$$

25. **(a)** The characteristic polynomial of the unforced equation is

$$s^2 + 9.$$

So the eigenvalues are $s = \pm 3i$, and the general solution of the unforced equation is

$$k_1 \cos 3t + k_2 \sin 3t.$$

To find one solution of the forced equation, we guess $y_p(t) = ke^{-t}$. Substituting $y_p(t)$ into the left-hand side of the differential equation, we obtain

$$\frac{d^2 y_p}{dt^2} + 9y_p = ke^{-t} + 9ke^{-t}$$

$$= 10ke^{-t}.$$

Hence, $k = 1/10$ yields a solution of the forced equation. The general solution of the forced equation is

$$y(t) = k_1 \cos 3t + k_2 \sin 3t + \tfrac{1}{10} e^{-t}.$$

(b) To find the solution satisfying the initial conditions $y(0) = y'(0) = 0$, we compute the derivative of the general solution

$$y'(t) = -3k_1 \sin 3t + 3k_2 \cos 3t - \tfrac{1}{10} e^{-t}.$$

Using the initial conditions and evaluating $y(t)$ and $y'(t)$ at $t = 0$, we obtain the simultaneous equations

$$\begin{cases} k_1 + \tfrac{1}{10} = 0 \\ 3k_2 - \tfrac{1}{10} = 0. \end{cases}$$

Solving for k_1 and k_2 gives $k_1 = -1/10$ and $k_2 = 1/30$. The solution of the initial-value problem is

$$y(t) = -\tfrac{1}{10} \cos 3t + \tfrac{1}{30} \sin 3t + \tfrac{1}{10} e^{-t}.$$

(c) Since the function $e^{-t}/10 \to 0$ quickly, the solution quickly approaches a solution of the unforced oscillator.

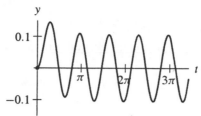

27. **(a)** The characteristic polynomial of the unforced equation is

$$s^2 + 2.$$

So the eigenvalues are $s = \pm i\sqrt{2}$, and the general solution of the unforced equation is

$$k_1 \cos \sqrt{2}\, t + k_2 \sin \sqrt{2}\, t.$$

To find one solution of the forced equation, we guess $y_p(t) = k$. Substituting into the left-hand side of the differential equation, we obtain

$$\frac{d^2 y_p}{dt^2} + 2 y_p = 0 + 2k$$

$$= 2k.$$

Hence, $k = -3/2$ yields a solution of the forced equation. The general solution of the forced equation is

$$y(t) = k_1 \cos \sqrt{2}\, t + k_2 \sin \sqrt{2}\, t - \tfrac{3}{2}.$$

(b) To find the solution satisfying the initial conditions $y(0) = y'(0) = 0$, we compute the derivative of the general solution

$$y'(t) = -\sqrt{2}\, k_1 \sin \sqrt{2}\, t + \sqrt{2}\, k_2 \cos \sqrt{2}\, t.$$

Using the initial conditions and evaluating $y(t)$ and $y'(t)$ at $t = 0$, we obtain the simultaneous equations

$$\begin{cases} k_1 - \tfrac{3}{2} = 0 \\ \sqrt{2}\, k_2 = 0. \end{cases}$$

Solving for k_1 and k_2 gives $k_1 = 3/2$ and $k_2 = 0$. The solution of the initial-value problem is

$$y(t) = \tfrac{3}{2} \cos \sqrt{2}\, t - \tfrac{3}{2}.$$

(c) The solution oscillates about the constant $y = -3/2$ with oscillations of amplitude $3/2$.

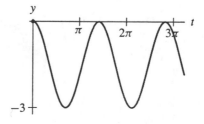

29. **(a)** The characteristic polynomial of the unforced equation is

$$s^2 + 9.$$

So the eigenvalues are $s = \pm 3i$, and the general solution of the unforced equation is

$$k_1 \cos 3t + k_2 \sin 3t.$$

To find one solution of the forced equation, we guess $y_p(t) = k$, where k is a constant. Substituting this guess into the left-hand side of the differential equation, we obtain

$$\frac{d^2 y_p}{dt^2} + 9y_p = 9k.$$

Hence, $k = 2/3$ yields a solution of the forced equation. The general solution of the forced equation is

$$y(t) = k_1 \cos 3t + k_2 \sin 3t + \tfrac{2}{3}.$$

(b) To find the solution satisfying the initial conditions $y(0) = y'(0) = 0$, we compute the derivative of the general solution

$$y'(t) = -3k_1 \sin 3t + 3k_2 \cos 3t.$$

Using the initial conditions and evaluating $y(t)$ and $y'(t)$ at $t = 0$, we obtain the simultaneous equations

$$\begin{cases} k_1 + \tfrac{2}{3} = 0 \\ 3k_2 = 0. \end{cases}$$

Solving for k_1 and k_2 gives $k_1 = -2/3$ and $k_2 = 0$. The solution of the initial-value problem is

$$y(t) = -\tfrac{2}{3} \cos 3t + \tfrac{2}{3}.$$

(c) The solution oscillates about the constant function $y = 2/3$ with amplitude $2/3$.

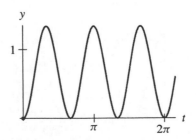

31. **(a)** The general solution for the homogeneous equation is

$$k_1 \cos 2t + k_2 \sin 2t.$$

Suppose $y_p(t) = at^2 + bt + c$. Substituting $y_p(t)$ into the differential equation, we get

$$\frac{d^2 y_p}{dt^2} + 4y_p = -3t^2 + 2t + 3$$

$$2a + 4(at^2 + bt + c) = -3t^2 + 2t + 3$$

$$4at^2 + 4bt + (2a + 4c) = -3t^2 + 2t + 3.$$

Therefore, $y_p(t)$ is a solution if and only if

$$\begin{cases} 4a = -3 \\ 4b = 2 \\ 2a + 4c = 3. \end{cases}$$

Therefore, $a = -3/4$, $b = 1/2$, and $c = 9/8$. The general solution is

$$y(t) = k_1 \cos 2t + k_2 \sin 2t - \tfrac{3}{4}t^2 + \tfrac{1}{2}t + \tfrac{9}{8}.$$

(b) To solve the initial-value problem, we use the initial conditions $y(0) = 2$ and $y'(0) = 0$ along with the general solution to form the simultaneous equations

$$\begin{cases} k_1 + \tfrac{9}{8} = 2 \\ 2k_2 + \tfrac{1}{2} = 0. \end{cases}$$

Therefore, $k_1 = 7/8$ and $k_2 = -1/4$. The solution is

$$y(t) = \tfrac{7}{8} \cos 2t - \tfrac{1}{4} \sin 2t - \tfrac{3}{4}t^2 + \tfrac{1}{2}t + \tfrac{9}{8}.$$

33. **(a)** For the unforced equation, the general solution is

$$k_1 \cos 2t + k_2 \sin 2t.$$

To find a particular solution of the forced equation, we guess $y_p(t) = at + b$. Substituting this guess into the differential equation, we get

$$\frac{d^2 y_p}{dt} + 4y_p = 3t + 2$$

$$0 + 4(at + b) = 3t + 2$$

$$4at + 4b = 3t + 2.$$

Therefore, $a = 3/4$ and $b = 1/2$ yield a solution. The general solution for the forced equation is

$$y(t) = k_1 \cos 2t + k_2 \sin 2t + \tfrac{3}{4}t + \tfrac{1}{2}.$$

(b) To solve the initial-value problem, we compute

$$y'(t) = -2k_1 \sin 2t + 2k_2 \cos 2t + \tfrac{3}{4}.$$

Evaluating $y(t)$ and $y'(t)$ at $t = 0$ and using the initial conditions, we obtain the simultaneous equations

$$\begin{cases} k_1 + \tfrac{1}{2} = 0 \\ 2k_2 + \tfrac{3}{4} = 0. \end{cases}$$

Hence, $k_1 = -1/2$ and $k_2 = -3/8$ provide the desired solution

$$y(t) = -\tfrac{1}{2}\cos 2t - \tfrac{3}{8}\cos 2t + \tfrac{3}{4}t + \tfrac{1}{2}.$$

(c) The solution tends to ∞ as it oscillates about the line $y = \tfrac{3}{4}t + \tfrac{1}{2}$.

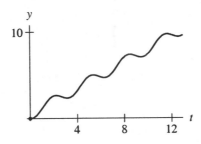

35. **(a)** The general solution of the homogeneous equation is

$$k_1 \cos 2t + k_2 \sin 2t.$$

To find a particular solution to the nonhomogeneous equation, we guess

$$y_p(t) = at^2 + bt + c.$$

Substituting $y_p(t)$ into the differential equation yields

$$\frac{d^2 y_p}{dt^2} + 4y_p = t - \frac{t}{20}$$

$$2a + 4(at^2 + bt + c) = t - \frac{t}{20}$$

$$(4a)t^2 + (4b)t + (2a + 4c) = t - \frac{t}{20}.$$

Equating coefficients, we obtain the simultaneous equations

$$\begin{cases} 4a = -\tfrac{1}{20} \\ 4b = 1 \\ 2a + 4c = 0. \end{cases}$$

Therefore, $a = -1/80$, $b = 1/4$, and $c = 1/160$ yield a solution to the nonhomogeneous equation, and the general solution of the nonhomogeneous equation is

$$y(t) = k_1 \cos 2t + k_2 \sin 2t - \tfrac{1}{80}t^2 + \tfrac{1}{4}t + \tfrac{1}{160}.$$

(b) To solve the initial-value problem with $y(0) = 0$ and $y'(0) = 0$, we have

$$\begin{cases} k_1 + \tfrac{1}{160} = 0 \\ 2k_2 + \tfrac{1}{4} = 0. \end{cases}$$

Therefore, $k_1 = -1/160$ and $k_2 = -1/8$, and the solution is

$$y(t) = -\tfrac{1}{160}\cos 2t - \tfrac{1}{8}\sin 2t - \tfrac{1}{80}t^2 + \tfrac{1}{4}t + \tfrac{1}{160}.$$

(c) Since the solution to the homogeneous equation is periodic with a small amplitude and since the solution to the nonhomogeneous equation goes to $-\infty$ at a rate determined by $-t^2/80$, the solution tends to $-\infty$.

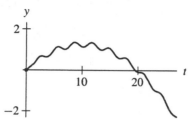

37. **(a)** We must find a particular solution. Using the result of Exercise 36, we guess $y_p(t) = ae^{-t} + b$, where a and b are constants to be determined. (We could solve two separate problems and add the answers, but this approach is more efficient.) Hence we have $dy_p/dt = -ae^{-t}$ and $d^2y_p/dt^2 = ae^{-t}$. Substituting these derivatives into the differential equation, we obtain

$$(a - 5a + 6a)e^{-t} + 6b = e^{-t} + 4,$$

which is satisfied if $2a = 1$ and $6b = 4$. Hence, $a = 1/2$ and $b = 2/3$ yield the particular solution $y_p(t) = e^{-t}/2 + 2/3$.
The general solution of the homogeneous equation is obtained from the characteristic polynomial

$$s^2 + 5s + 6,$$

whose roots are $s = -2$ and $s = -3$.
Hence the general solution is

$$y(t) = k_1 e^{-2t} + k_2 e^{-3t} + \tfrac{1}{2}e^{-t} + \tfrac{2}{3}.$$

(b) To obtain the solution to the initial-value problem specified, we note that

$$y(0) = k_1 + k_2 + 1/2 + 2/3 \quad \text{and} \quad y'(0) = -2k_1 - 3k_2 - 1/2.$$

Using the initial conditions $y(0) = 0$ and $y'(0) = 0$, we have $k_1 = -3$ and $k_2 = 11/6$. The solution is

$$y(t) = -3e^{-2t} + \tfrac{11}{6}e^{-3t} + \tfrac{1}{2}e^{-t} + \tfrac{2}{3}.$$

(c) All of the exponential terms in the solution to the initial-value problem tend to 0. Hence, the solution tends to the constant $y = 2/3$. The rate that this solution tends to the constant is determined by $e^{-t}/2$, which is the largest of the terms that tend to zero when t is large.

39. **(a)** First, to find a particular solution of the forced equation, we guess

$$y_p(t) = at + b + ce^{-t}.$$

For y_p, $dy_p/dt = a - ce^{-t}$ and $d^2y_p/dt^2 = ce^{-t}$. Substituting these derivatives into the differential equation and collecting terms gives

$$(c - 6c + 8c)e^{-t} + (8a)t + (6a + 8b) = 2t + e^{-t},$$

which holds if $3c = 1$, $8a = 2$, and $6a + 8b = 0$. Hence, $c = 1/3$, $a = 1/4$, and $b = -3/16$ yield the solution

$$y_p(t) = -\tfrac{3}{16} + \tfrac{1}{4}t + \tfrac{1}{3}e^{-t}.$$

The characteristic polynomial of the homogeneous equation is

$$s^2 + 6s + 8,$$

which has roots $s = -4$ and $s = -2$, so the general solution of the forced equation is

$$y(t) = k_1 e^{-4t} + k_2 e^{-2t} - \tfrac{3}{16} + \tfrac{1}{4}t + \tfrac{1}{3}e^{-t}.$$

(b) To find the solution for the initial conditions $y(0) = 0$ and $y'(0) = 0$, we solve

$$\begin{cases} k_1 + k_2 - \tfrac{3}{16} + \tfrac{1}{3} = 0 \\ -4k_1 - 2k_2 + \tfrac{1}{4} - \tfrac{1}{3} = 0. \end{cases}$$

Thus, $k_1 = 5/48$ and $k_2 = -1/4$ yield the solution

$$y(t) = \tfrac{5}{48}e^{-4t} - \tfrac{1}{4}e^{-2t} - \tfrac{3}{16} + \tfrac{1}{4}t + \tfrac{1}{3}e^{-t}$$

of the initial-value problem.

(c) All exponential terms in the solution tend to zero. Hence, the solution tends to infinity linearly in t and is close to $t/4$ for large t.

41. **(a)** To find the general solution, we first guess

$$y_p(t) = ae^{-t} + bt + c,$$

where a, b and c are constants to be determined. For y_p,

$$\frac{dy_p}{dt} = -ae^{-t} + b \quad \text{and} \quad \frac{d^2y_p}{dt^2} = ae^{-t}.$$

Substituting these derivatives into the differential equation and collecting terms gives

$$(a + 4a)e^{-t} + (4b)t + (4c) = t + e^{-t},$$

which is satisfied if $5a = 1$, $4b = 1$, and $4c = 0$. Hence, a solution is

$$y_p(t) = \tfrac{1}{5}e^{-t} + \tfrac{1}{4}t.$$

To find the general solution of the homogeneous equation, we note that the characteristic polynomial $s^2 + 4$ has roots $s = \pm 2i$. Hence, the general solution of the forced equation is

$$y(t) = k_1 \cos 2t + k_2 \sin 2t + \tfrac{1}{5}e^{-t} + \tfrac{1}{4}t.$$

(b) To find the solution with the desired initial conditions, we note that $y(0) = k_1 + 0 + 1/5$ and $y'(0) = 2k_2 - 1/5 + 1/4$. We must solve the simultaneous equations

$$\begin{cases} k_1 + \frac{1}{5} = 0 \\ 2k_2 + \frac{1}{20} = 0. \end{cases}$$

Thus, $k_1 = -1/5$ and $k_2 = -1/40$ yield the solution

$$y(t) = -\tfrac{1}{5}\cos 2t - \tfrac{1}{40}\sin 2t + \tfrac{1}{5}e^{-t} + \tfrac{1}{4}t.$$

(c) Since all of the terms in the solution except $t/4$ are bounded for $t > 0$, the solution tends to infinity at a rate that is determined by $t/4$.

EXERCISES FOR SECTION 4.2

1. Recalling that the real part of e^{it} is $\cos t$, we see that the complex version of this equation is

$$\frac{d^2y}{dt^2} + 3\frac{dy}{dt} + 2y = e^{it}.$$

To find a particular solution, we guess $y_c(t) = ae^{it}$. Then $dy_c/dt = iae^{it}$ and $d^2y_c/dt^2 = -ae^{it}$. Substituting these derivatives into the equation and collecting terms yields

$$(-a + 3ia + 2a)e^{-it} = e^{it},$$

which is satisfied if

$$(1 + 3i)a = 1.$$

Hence, we must have

$$a = \frac{1}{1 + 3i} = \frac{1}{10} - \frac{3}{10}i.$$

So

$$y_c(t) = \frac{1 - 3i}{10}e^{it} = \frac{1 - 3i}{10}(\cos t + i\sin t)$$

is a particular solution of the complex version of the equation. Taking the real part, we obtain the solution

$$y(t) = \tfrac{1}{10}\cos t + \tfrac{3}{10}\sin t.$$

To produce the general solution of the homogeneous equation, we note that the characteristic polynomial $s^2 + 3s + 2$ has roots $s = -2$ and $s = -1$. So the general solution is

$$y(t) = k_1 e^{-2t} + k_2 e^{-t} + \tfrac{1}{10}\cos t + \tfrac{3}{10}\sin t.$$

3. Recalling that the imaginary part of e^{it} is $\sin t$, the complex version of the equation is

$$\frac{d^2y}{dt^2} + 3\frac{dy}{dt} + 2y = e^{it}.$$

This equation is precisely the same complex equation as in Exercise 1. Hence, we have already computed the solution

$$y_c(t) = \frac{1-3i}{10}(\cos t + i\sin t).$$

In this case we take the imaginary part

$$y(t) = -\tfrac{3}{10}\cos t + \tfrac{1}{10}\sin t$$

to obtain a solution of the original differential equation.

The general solution of the homogeneous equation is the same as in Exercise 1, so the general solution is

$$y(t) = k_1 e^{-2t} + k_2 e^{-t} - \tfrac{3}{10}\cos t + \tfrac{1}{10}\sin t.$$

5. The complex version of this equation is

$$\frac{d^2y}{dt^2} + 6\frac{dy}{dt} + 8y = e^{it}.$$

We guess a particular solution of the form $y_c(t) = ae^{it}$. Then $dy_c/dt = iae^{it}$ and $d^2y/dt^2 = -ae^{it}$. Substituting these derivatives into the complex differential equation yields

$$(-a + 6ia + 8a)e^{it} = e^{it},$$

which is satisfied if $(7+6i)a = 1$. Then $a = 1/(7+6i)$, and

$$y_c(t) = \frac{7-6i}{85}e^{it} = \frac{7-6i}{85}(\cos t + i\sin t).$$

The real part

$$y(t) = \tfrac{7}{85}\cos t + \tfrac{6}{85}\sin t$$

is a solution of the original differential equation.

To find the general solution of the homogeneous equation, we note that the characteristic polynomial $s^2 + 6s + 8$ has roots $s = -4$ and $s = -2$. Consequently, the general solution of the original nonhomogeneous equation is

$$y(t) = k_1 e^{-4t} + k_2 e^{-2t} + \tfrac{7}{85}\cos t + \tfrac{6}{85}\sin t.$$

7. The complex version of this equation is

$$\frac{d^2y}{dt^2} + 4\frac{dy}{dt} + 13y = 3e^{2it},$$

so we guess $y_c(t) = ae^{2it}$ to find a particular solution. Substituting $y_c(t)$ into the differential equation gives

$$(-4a + 8ai + 13a)e^{2it} = 3e^{2it},$$

which is satisfied if $(9 + 8i)a = 3$. Thus, $y_c(t)$ is a solution if

$$y_c(t) = \frac{3}{9 + 8i} e^{2it} = \frac{27 - 24i}{145}(\cos 2t + i \sin 2t).$$

A particular solution of the original equation is the real part

$$y(t) = \tfrac{27}{145} \cos 2t + \tfrac{24}{145} \sin 2t.$$

To find the general solution of the homogeneous equation, we note that the characteristic polynomial $s^2 + 4s + 13$ has roots $s = -2 \pm 3i$. Hence, the general solution of the original forced equation is

$$y(t) = k_1 e^{-2t} \cos 3t + k_2 e^{-2t} \sin 3t + \tfrac{27}{145} \cos 2t + \tfrac{24}{145} \sin 2t.$$

9. The complex version of this equation is

$$\frac{d^2y}{dt^2} + 4\frac{dy}{dt} + 20y = -3e^{2it},$$

and we guess that there is a solution of the form $y_c(t) = ae^{2it}$. Substituting this guess into the differential equation yields

$$(-4a + 8ia + 20a)e^{2it} = -3e^{2it},$$

which can be simplified to

$$(16 + 8i)ae^{2it} = -3e^{2it}.$$

Thus, $y_c(t)$ is a solution if $a = -3/(16 + 8i)$. We have

$$y_c(t) = \frac{-3}{16 + 8i} e^{2it} = \left(-\frac{3}{20} + \frac{3}{40}i\right)\left(\cos 2t + i \sin 2t\right).$$

We take the imaginary part to obtain a solution

$$y(t) = \tfrac{3}{40} \cos 2t - \tfrac{3}{20} \sin 2t$$

of the original equation.

To find the general solution of the homogeneous equation, we note that the characteristic polynomial is $s^2 + 4s + 20$, whose roots are $s = -2 \pm 4i$. Hence, the general solution of the original equation is

$$y(t) = k_2 e^{-2t} \cos 4t + k_2 e^{-2t} \sin 4t - \tfrac{3}{20} \sin 2t + \tfrac{3}{40} \cos 2t.$$

11. From Exercise 5, we know that the general solution of this equation is

$$y(t) = k_1 e^{-4t} + k_2 e^{-2t} + \tfrac{7}{85} \cos t + \tfrac{6}{85} \sin t.$$

To find the desired solution, we must solve for k_1 and k_2 using the initial conditions. We have

$$\begin{cases} k_1 + k_2 + \tfrac{7}{85} = 0 \\ -4k_1 - 2k_2 + \tfrac{6}{85} = 0. \end{cases}$$

We obtain $k_1 = 2/17$ and $k_2 = -1/5$. The desired solution is

$$y(t) = \tfrac{2}{17} e^{-4t} - \tfrac{1}{5} e^{-2t} + \tfrac{7}{85} \cos t + \tfrac{6}{85} \sin t.$$

13. From Exercise 9, we know that the general solution of this equation is

$$y(t) = k_1 e^{-2t} \cos 4t + k_2 e^{-2t} \sin 4t - \tfrac{3}{20} \sin 2t + \tfrac{3}{40} \cos 2t.$$

To find the desired solution, we must solve for k_1 and k_2 using the initial conditions. We have

$$\begin{cases} k_1 + \tfrac{3}{40} = 0 \\ -2k_1 + 4k_2 - \tfrac{6}{20} = 0. \end{cases}$$

We obtain $k_1 = -3/40$ and $k_2 = 3/80$. The desired solution is

$$y(t) = -\tfrac{3}{40} e^{-2t} \cos 4t + \tfrac{3}{80} e^{-2t} \sin 4t - \tfrac{3}{20} \sin 2t + \tfrac{3}{40} \cos 2t.$$

15. (a) If we guess

$$y_p(t) = a \cos 3t + b \sin 3t,$$

then

$$y_p'(t) = -3a \sin 3t + 3b \cos 3t$$

and

$$y_p''(t) = -9a \cos 3t - 9b \sin 3t.$$

Substituting this guess and its derivatives into the differential equation gives

$$(-8a + 9b) \cos 3t + (-9a - 8b) \sin 3t = \cos 3t.$$

Thus $y_p(t)$ is a solution if a and b satisfy the simultaneous equations

$$\begin{cases} -8a + 9b = 1 \\ -9a - 8b = 0. \end{cases}$$

Solving these equations for a and b, we obtain $a = -8/145$ and $b = 9/145$, so

$$y_p(t) = -\tfrac{8}{145} \cos 3t + \tfrac{9}{145} \sin 3t$$

is a solution.

(b) If we guess

$$y_p(t) = A\cos(3t + \phi),$$

then

$$y_p'(t) = -3A\sin(3t + \phi)$$

and

$$y_p''(t) = -9A\cos(3t + \phi).$$

Substituting this guess and its derivatives into the differential equation yields

$$-8A\cos(3t + \phi) - 9A\sin(3t + \phi) = \cos 3t.$$

Using the trigonometric identities for the sine and cosine of the sum of two angles, we have

$$-8A\,(\cos 3t\cos\phi - \sin 3t\sin\phi) - 9A\,(\sin 3t\cos\phi + \cos 3t\sin\phi) = \cos 3t.$$

This equation can be rewritten as

$$(-8A\cos\phi - 9A\sin\phi)\cos 3t + (8A\sin\phi - 9A\cos\phi)\sin 3t = \cos 3t.$$

It holds if

$$\begin{cases} -8A\cos\phi - 9A\sin\phi = 1 \\ 9A\cos\phi - 8A\sin\phi = 0. \end{cases}$$

Multiplying the first equation by 9 and the second by 8 and adding yields

$$145A\sin\phi = -9.$$

Similarly, multiplying the first equation by -8 and the second by 9 and adding yields

$$145A\cos\phi = -8.$$

Taking the ratio gives

$$\frac{\sin\phi}{\cos\phi} = \tan\phi = \frac{9}{8}.$$

Also, squaring both $145A\sin\phi = -9$ and $145A\cos\phi = -8$ yields

$$145^2 A^2\cos^2\phi + 145^2 A^2\sin^2\phi = 145,$$

so $A^2 = 1/145$.
We can use either $A = 1/\sqrt{145}$ or $A = -1/\sqrt{145}$, but this choice of sign for A effects the value of ϕ. If we pick $A = -1/\sqrt{145}$, then $\sqrt{145}\sin\phi = 9$, $\sqrt{145}\cos\phi = 8$, and $\tan\phi = 9/8$. In this case, $\phi = \arctan(9/8)$. Hence, a particular solution of the original equation is

$$y_p(t) = \frac{1}{\sqrt{145}}\cos\left(3t + \arctan\frac{9}{8}\right).$$

17. Due to the damping ($p > 0$), the solution of the homogeneous equation (the unforced response) tends to zero. Solutions of the nonhomogeneous equation (the forced response) are sums of terms involving $\cos t$ and $\sin t$. Hence they oscillate with period 2π. This observation eliminates Figures (i) and (iv).

Figures (ii) and (iii) differ in the amplitude of the oscillations of the forced response. We compute a particular solution of the equation as usual, considering the complex version

$$\frac{d^2y}{dt^2} + \frac{dy}{dt} + 3y = e^{it}.$$

We obtain a particular solution of the form $y_c(t) = ae^{it}$ by substituting $y_c(t)$ into the differential equation. We have

$$(2+i)ae^{it} = e^{it},$$

which is satisfied if $a = 1/(2+i)$. Since

$$y_c(t) = \frac{2-i}{5}\,(\cos t + i\sin t),$$

a particular solution of the forced equation is

$$y(t) = \tfrac{2}{5}\cos t + \tfrac{1}{5}\sin t.$$

Due to the damping, all solutions have the same long-term behavior including the long-term amplitude of the oscillations. Since the amplitude of $y(t)$ is $|a| = 1/\sqrt{5} \approx 0.44$, the graphs shown in Figure (iii) are not possible. Our analysis is consistent with the graphs shown in Figure (ii).

19. Due to the damping ($p > 0$), the solution of the homogeneous equation (the unforced response) tends to zero. Solutions of the nonhomogeneous equation (the forced response) are sums of terms involving $\cos 3t$ and $\sin 3t$. Hence they oscillate with period $2\pi/3$. This observation eliminates Figures (ii) and (iii).

The difference between the graphs in Figure (i) and the graphs in Figure (iv) is the rate at which solutions tend toward each other. This rate is the same as the rate at which the solutions of the homogeneous equation tend to zero. The characteristic polynomial of the homogeneous equation is $s^2 + s + 1$, so the eigenvalues are $s = (-1 \pm \sqrt{3}\,i)/2$. The rate that solutions tend to zero is determined by the real part of these complex eigenvalues. Hence, all solutions of the homogeneous equation tend to zero at the rate of $e^{-0.5t}$, which is definitely faster than the rate in Exercise 18. The graphs in Figure (iv) are consistent with this analysis.

21. By Exercise 5 we know that one solution of

$$\frac{d^2y}{dt^2} + 6\frac{dy}{dt} + 8y = \cos t$$

is

$$y_1(t) = \tfrac{7}{85}\cos t + \tfrac{6}{85}\sin t.$$

Using the result of Exercise 20, a particular solution of the given equation is $y_2(t) = 5y_1(t)$. In other words,

$$y_2(t) = \tfrac{7}{17}\cos t + \tfrac{6}{17}\sin t$$

is a particular solution to the equation in this exercise.

The general solution of the homogeneous equation is the same as in Exercise 5, so the general solution for this exercise is

$$y(t) = k_1 e^{-4t} + k_2 e^{-2t} + \tfrac{7}{17} \cos t + \tfrac{6}{17} \sin t.$$

23. **(a)** Using the fact that the real part of $e^{(-1+i)t}$ is $e^{-t} \cos t$, the complex version of this equation is

$$\frac{d^2 y}{dt^2} + 4\frac{dy}{dt} + 20y = e^{(-1+i)t}.$$

Guessing $y_c(t) = ae^{(-1+i)t}$ yields

$$a(-1+i)^2 e^{(-1+i)t} + 4a(-1+i)e^{(-1+i)t} + 20ae^{(-1+i)t} = e^{(-1+i)t}.$$

Simplifying we have

$$a(16 + 2i)e^{(-1+i)t} = e^{(-1+i)t}.$$

Thus, $y_c(t)$ is a solution of the complex differential equation if $a = 1/(16 + 2i)$, and we have

$$y_c(t) = \left(\tfrac{4}{65} - \tfrac{1}{130}i \right) e^{-t} (\cos t + i \sin t).$$

So one solution of the original equation is

$$y_p(t) = \tfrac{4}{65} e^{-t} \cos t + \tfrac{1}{130} e^{-t} \sin t.$$

To find the general solution of the homogeneous equation, we note that the characteristic polynomial $s^2 + 4s + 20$ has roots $s = -2 \pm 4i$.

Hence, the general solution of the original equation is

$$y(t) = k_1 e^{-2t} \cos 4t + k_1 e^{-2t} \sin 4t + \tfrac{4}{65} e^{-t} \cos t + \tfrac{1}{130} e^{-t} \sin t.$$

(b) All four terms in the general solution tend to zero as $t \to \infty$. Hence, all solutions tend to zero as $t \to \infty$. The terms with factors of e^{-2t} tend to zero very quickly, which leaves the terms of the particular solution $y_p(t)$ as the largest terms, so all solutions are asymptotic to $y_p(t)$. Since the solution $y_p(t)$ oscillates with period 2π and the amplitude of its oscillations decreases at the rate of e^{-t}, all solutions oscillate with this period and decaying amplitude.

25. Note that the real part of

$$(a - bi)(\cos \omega t + i \sin \omega t)$$

is $g(t)$. Hence, we must find k and ϕ such that

$$ke^{i\phi} = a - bi.$$

Using the polar form of the complex number $z = a - bi$, we see that

$$ke^{i\phi} = a - bi = z = |z|e^{i\theta},$$

where θ is the polar angle for z (see Appendix B). Therefore, we can choose

$$k = |z| = \sqrt{a^2 + b^2} \quad \text{and} \quad \phi = \theta.$$

27. Note that the real part of

$$(k_1 - ik_2)e^{i\beta t} = (k_1 - ik_2)(\cos \beta t + i \sin \beta t)$$

is

$$y(t) = k_1 \cos \beta t + k_2 \sin \beta t.$$

Let $Ke^{i\phi}$ be the polar form of the complex number $k_1 + ik_2$. Then the polar form of $k_1 - ik_2$ is $Ke^{-i\phi}$. Using the Laws of Exponents and Euler's formula, we have

$$\begin{aligned}
(k_1 - ik_2)e^{i\beta t} &= Ke^{-i\phi}e^{i\beta t} \\
&= Ke^{i(\beta t - \phi)} \\
&= K(\cos(\beta t - \phi) + i \sin(\beta t - \phi),
\end{aligned}$$

and the real part is $K \cos(\beta t - \phi)$. Hence, we see that

$$y(t) = k_1 \cos \beta t + k_2 \sin \beta t$$

can be rewritten as

$$y(t) = K \cos(\beta t - \phi).$$

EXERCISES FOR SECTION 4.3

1. The complex version of this equation is

$$\frac{d^2 y}{dt^2} + 9y = e^{it}.$$

Guessing $y_c(t) = ae^{it}$ as a particular solution and substituting this guess into the left-hand side of the differential equation yields

$$8ae^{it} = e^{it}.$$

Thus, $y_c(t)$ is a solution if $8a = 1$. The real part of

$$y_c(t) = \tfrac{1}{8}e^{it} = \tfrac{1}{8}(\cos t + i \sin t)$$

is $y(t) = \tfrac{1}{8} \cos t$. This $y(t)$ is a solution to the original differential equation. [Because there is no dy/dt-term (no damping), we could have guessed a solution of the form $y(t) = a \cos t$ instead of using the complex version of the equation.]

To find the general solution of the homogeneous equation, we note that the characteristic polynomial is $s^2 + 9$, which has roots $s = \pm 3i$. So the general solution of the original equation is

$$y(t) = k_1 \cos 3t + k_2 \sin 3t + \tfrac{1}{8} \cos t.$$

3. The complex version of this equation is

$$\frac{d^2y}{dt^2} + 4y = -e^{it/2}.$$

Guessing $y_c(t) = ae^{it/2}$ as a particular solution and substituting into the equation yields

$$\tfrac{15}{4}ae^{it/2} = -e^{it/2}.$$

Thus, $y_c(t)$ is a solution if $\tfrac{15}{4}a = -1$. The real part of

$$y_c(t) = -\frac{4}{15}e^{it/2} = -\frac{4}{15}\left(\cos\frac{t}{2} + i\sin\frac{t}{2}\right)$$

is

$$y(t) = -\frac{4}{15}\cos\frac{t}{2}.$$

This $y(t)$ is a solution to the original differential equation. [Because there is no dy/dt-term (no damping), we could have guessed a solution of the form $y(t) = a\cos t/2$ instead of using the complex version of the equation.]

To find the general solution of the homogeneous equation, we note that the characteristic polynomial is $s^2 + 4$, which has roots $s = \pm 2i$. So the general solution of the original equation is

$$y(t) = k_1\cos 2t + k_2\sin 2t - \frac{4}{15}\cos\frac{t}{2}.$$

5. The complex version of the equation is

$$\frac{d^2y}{dt^2} + 9y = 2e^{3it}.$$

Guessing $y_c(t) = ae^{3it}$ as a particular solution and substituting this guess into the left-hand side of the differential equation, we see that a must satisfy

$$-9a + 9a = 2,$$

which is impossible. Hence, the forcing is in resonance with the associated homogeneous equation. We must make a second guess of $y_c(t) = ate^{3it}$. This guess gives

$$y'_c(t) = a(1 + 3it)e^{3it}$$

and

$$y''_c(t) = a(6i - 9t)e^{3it}.$$

Substituting $y_c(t)$ and its second derivative into the differential equation, we obtain

$$a(6i - 9t)e^{3it} + 9ate^{3it} = 2e^{3it},$$

which simplifies to

$$6aie^{3it} = 2e^{3it}.$$

Thus, $y_c(t)$ is a solution if $a = 2/(6i) = -i/3$. Taking the real part of

$$y_c(t) = -\tfrac{1}{3}it(\cos 3t + i \sin 3t),$$

we obtain the solution

$$y(t) = \tfrac{1}{3}t \sin 3t$$

of the original equation.

To find the general solution of the homogeneous equation, we note that the characteristic polynomial is $s^2 + 9$, which has roots $s = \pm 3i$.

Hence, the general solution of the original equation is

$$y(t) = k_1 \cos 3t + k_2 \sin 3t + \tfrac{1}{3}t \sin 3t.$$

7. The complex version of this equation is

$$\frac{d^2 y}{dt^2} + 3y = e^{3it}.$$

Guessing $y_c(t) = ae^{3it}$ as a particular solution and substituting this guess into the left-hand side of the differential equation yields

$$-6ae^{3it} = e^{3it}.$$

Thus, $y_c(t)$ is a solution if $-6a = 1$. The real part of

$$y_c(t) = -\tfrac{1}{6}e^{3it} = -\tfrac{1}{6}(\cos 3t + i \sin 3t)$$

is $y(t) = -\tfrac{1}{6}\cos 3t$. This $y(t)$ is a solution to the original differential equation. [Because there is no dy/dt-term (no damping), we could have guessed a solution of the form $y(t) = a \cos 3t$ instead of using the complex version of the equation.]

To find the general solution of the homogeneous equation, we note that the characteristic polynomial is $s^2 + 3$, which has roots $s = \pm\sqrt{3}\,i$.

Hence, the general solution of the original equation is

$$y(t) = k_1 \cos \sqrt{3}\,t + k_2 \sin \sqrt{3}\,t - \tfrac{1}{6} \cos 3t.$$

9. From Exercise 1, we know that the general solution is

$$y(t) = k_1 \cos 3t + k_2 \sin 3t + \tfrac{1}{8} \cos t.$$

So

$$y'(t) = -3k_1 \sin 3t + 3k_2 \cos 3t - \tfrac{1}{8} \sin t.$$

Using the initial conditions $y(0) = 0$ and $y'(0) = 0$, we obtain the simultaneous equations

$$\begin{cases} k_1 + \tfrac{1}{8} = 0 \\ 3k_2 = 0, \end{cases}$$

which imply that $k_1 = -1/8$ and $k_2 = 0$. The solution to the initial-value problem is

$$y(t) = -\tfrac{1}{8} \cos 3t + \tfrac{1}{8} \cos t.$$

11. First we find the general solution by considering the complex version of the equation

$$\frac{d^2y}{dt^2} + 5y = 3e^{2it}.$$

Guessing a particular solution of the form $y_c(t) = ae^{2it}$ and substituting this guess into the left-hand side of the equation yields

$$ae^{2it} = 3e^{2it}.$$

Thus, $y_c(t)$ is a solution if $a = 3$. The real part of

$$y_c(t) = 3(\cos 2t + i \sin 2t)$$

is $y(t) = 3 \cos 2t$. This $y(t)$ is a solution to the original differential equation. [Because there is no dy/dt-term (no damping), we could have guessed a solution of the form $y(t) = a \cos 2t$ instead of using the complex version of the equation.]

To find the general solution of the homogeneous equation, we note that the characteristic polynomial is $s^2 + 5$, which has roots $s = \pm\sqrt{5}\,i$. Hence, the general solution is

$$y(t) = k_1 \cos \sqrt{5}\,t + k_2 \sin \sqrt{5}\,t + 3 \cos 2t.$$

Note that

$$y'(t) = -\sqrt{5}\,k_1 \sin \sqrt{5}\,t + \sqrt{5}\,k_2 \cos \sqrt{5}\,t - 6 \sin 2t.$$

Using the initial conditions $y(0) = 0$ and $y'(0) = 0$, we obtain the simultaneous equations

$$\begin{cases} k_1 + 3 = 0 \\ \sqrt{5}\,k_2 = 0, \end{cases}$$

which imply that $k_1 = -3$ and $k_2 = 0$. The solution to the initial-value problem is

$$y(t) = -3 \cos \sqrt{5}\,t + 3 \cos 2t.$$

13. From Exercise 5, we know that the general solution is

$$y(t) = k_1 \cos 3t + k_2 \sin 3t + \tfrac{1}{3}t \sin 3t.$$

So

$$y'(t) = -3k_1 \sin 3t + 3k_2 \cos 3t + \tfrac{1}{3} \sin 3t + t \cos 3t.$$

From the initial condition $y(0) = 2$, we see that $k_1 = 2$. Using the initial condition $y'(0) = -9$, we have $3k_2 = -9$. Hence, $k_2 = -3$. The solution to the initial-value problem is

$$y(t) = 2 \cos 3t - 3 \sin 3t + \tfrac{1}{3}t \sin 3t.$$

15. The characteristic polynomial of the unforced equation is $s^2 + 4$, which has roots $s = \pm 2i$. So the natural frequency is $2/(2\pi)$, and the forcing frequency is $9/(8\pi)$.

    (a) The frequency of the beats is

$$\frac{\frac{9}{4} - 2}{4\pi} = \frac{1}{16\pi}.$$

    (b) The frequency of the rapid period oscillations is

$$\frac{\frac{9}{4} + 2}{4\pi} = \frac{17}{16\pi}.$$

    (c)

17. The characteristic polynomial of the unforced equation is $s^2 + 5$, which has roots $s = \pm i\sqrt{5}$. So the natural frequency is $\sqrt{5}/(2\pi)$, and the forcing frequency is $2/(2\pi)$.

    (a) The frequency of the beats is

$$\frac{\sqrt{5} - 2}{4\pi}.$$

    (b) The frequency of the rapid oscillations is

$$\frac{\sqrt{5} + 2}{4\pi}.$$

    (c)

19.   **(a)** To find the general solution, we deal with each of the forcing terms separately. In other words, we find solutions to

$$\frac{d^2y}{dt^2} + 15y = \cos 4t \quad \text{and} \quad \frac{d^2y}{dt^2} + 15y = 2\sin t$$

    separately and add them to get a solution to the original equation (see Exercise 36 in Section 4.1).

First consider the equation whose forcing term is $\cos 4t$. The complex version of the equation is

$$\frac{d^2y}{dt^2} + 15y = e^{4it}.$$

We guess a solution of the form $y_c(t) = ae^{4it}$ and substitute it into the differential equation to obtain

$$a(-16 + 15)e^{4it} = e^{4it},$$

which is satisfied if $a = -1$. Hence, by taking the real part of

$$y_c(t) = -(\cos 4t + i\sin 4t),$$

we obtain the solution $y_1(t) - -\cos 4t$.

Similarly, to find a solution of the equation whose forcing term is $2\sin t$, we consider

$$\frac{d^2y}{dt^2} + 15y = 2e^{it}$$

and guess $y_c(t) = be^{it}$. Substituting this guess into the equation yields

$$b(-1 + 15)e^{it} = 2e^{it},$$

which is satisfied if $b = 1/7$. So, by taking the imaginary part of

$$y_c(t) = \tfrac{1}{7}(\cos t + i\sin t),$$

we obtain the solution $y_2(t) = \tfrac{1}{7}\sin t$.

To obtain the general solution of the unforced equation, we note that the characteristic polynomial is $s^2 + 15$, which has roots $s = \pm i\sqrt{15}$. So the general solution of the original differential equation is

$$y(t) = k_1\cos\sqrt{15}\,t + k_2\sin\sqrt{15}\,t - \cos 4t + \tfrac{1}{7}\sin t.$$

(b) To obtain the initial conditions $y(0) = 0$ and $y'(0) = 0$, we note that

$$y'(t) = -\sqrt{15}\,k_1\sin\sqrt{15}\,t + \sqrt{15}\,k_2\cos\sqrt{15}\,t + 4\sin 4t + \tfrac{1}{7}\cos t.$$

Hence, we must solve the simultaneous equations

$$\begin{cases} k_1 - 1 = 0 \\ \sqrt{15}\,k_2 + \tfrac{1}{7} = 0 \end{cases}$$

for k_1 and k_2. We obtain $k_1 = 1$ and $k_2 = -1/(7\sqrt{15}) = -\sqrt{15}/105$. Hence, the solution to the initial value problem is

$$y(t) = \cos\sqrt{15}\,t - \frac{\sqrt{15}}{105}\sin\sqrt{15}\,t - \cos 4t + \frac{1}{7}\sin t.$$

(c)

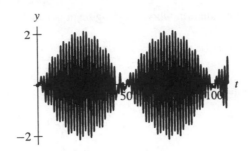

(d) The solution is the sum of the general solution of the unforced equation and particular solutions of the forced equations with each forcing term considered separately. Since the forcing function $\cos 4t$ is so close to resonance, we expect to see large amplitude beats with frequency $(\sqrt{15} - 4)/(4\pi)$ and rapid oscillations with frequency $(\sqrt{15} + 4)/(4\pi)$. The particular solution for the $2 \sin t$ forcing term has a relatively small amplitude, and therefore it will not have a significant effect on the final oscillations most of the time.

21. The graph has rapid oscillations with a period that is approximately 2 units of t, and the period of the beats (slow oscillations) is approximately 12 units of t (slightly less than one-half of a period of a beat is shown). Since equation (iii) is resonant and equation (iv) has no beats, the only two possibilities are equation (i) and equation (ii). Since difference $\sqrt{15} - 4$ (equation (i)) is much less than the difference $\sqrt{16} - 3$ (equation (ii)), the period of the slow oscillations for equation (ii) is much less than the period of the slow oscillations for equation (i). Hence, this graph corresponds to equation (ii).

23. The graph has beats, so it cannot correspond to equation (iii), which is resonant, or equation (iv), which has constant forcing. This equation has very long period beats, so the natural and forcing frequencies must be close. Because $\sqrt{15}$ is much closer to 4 than 3 is to 4, this graph corresponds to equation (i). (See the solution to Exercise 21.)

25. The frequency of the stomping is almost the same as the natural frequency of the swaying motion of the stadium. Therefore, the stadium structure reacted violently due to the resonant effects of the stomping.

27. The equation of motion for the unforced mass-spring system is

$$\frac{d^2y}{dt^2} + 16y = 0,$$

so the natural period is $2\pi/4 = \pi/2 \approx 1.57$.

Tapping with the hammer as shown increases the velocity if the mass is moving to the right at the time of the tap and decreases the velocity if the mass is moving to the left at the time of the tap. Faster motion results in higher amplitude oscillations. Since none of the tapping periods is exactly $\pi/2$, the taps sometimes increase the amplitude and sometimes decrease the amplitude of the oscillations (that is, resonance does not occur).

The period $T = 3/2$ is closest to the natural period and hence for taps with this period we expect the largest amplitude oscillations.

EXERCISES FOR SECTION 4.4

1. Rubbing a finger around the edge of the glass starts the glass vibrating. The finger then skips along the glass, giving a forcing term which has the same frequency as the natural frequency of vibration of the glass. Pressing harder changes the size of the forcing. Since the frequency is determined by the motion of the glass itself, pushing harder does not alter the frequency.

3. Given that

$$y_p''(t) + p y_p'(t) + q y_p(t) = g(t)$$

for all t, we have

$$y_p''(t + \theta) + p y_p'(t + \theta) + q y_p(t + \theta) = g(t + \theta).$$

Let $z(t) = y_p(t + \theta)$. Then

$$z'(t) = y_p'(t + \theta) \quad \text{and} \quad z''(t) = y_p''(t + \theta)$$

by the Chain Rule. Thus,

$$z''(t) + p z'(t) + q z(t) = g(t + \theta),$$

and $z(t) = y_p(t + \theta)$ is a solution of

$$\frac{d^2 y}{dt^2} + p \frac{dy}{dt} + qy = g(t + \theta).$$

5. As in Exercise 4, the forced response is

$$y(t) = A_1 \cos \omega_1 t + A_2 \cos \omega_2 t.$$

where

$$A_i = \frac{1}{\sqrt{(q - \omega_i^2)^2 + p^2 \omega_i^2}}$$

for $i = 1, 2$.

(a) Setting $q = \omega_1^2$, we have

$$\frac{A_1}{A_2} = \frac{\sqrt{(\omega_1^2 - \omega_2^2)^2 + p^2 \omega_2^2}}{p \omega_1}.$$

(b) Let $R(p) = A_1/A_2$. We see that $R(p) \to \infty$ like $1/p$ as $p \to 0$ and $R(p) \to 0$ like $1/p$ as $p \to \infty$. Moreover

$$\frac{dR}{dp} = \frac{-(\omega_1^2 - \omega_2^2)^2}{p^2 \omega_1^2 \sqrt{(\omega_1^2 - \omega_2^2)^2 + p^2 \omega_2^2}},$$

which is negative for all $p > 0$. Hence, $R(p)$ is monotonically decreasing.

7. (a)

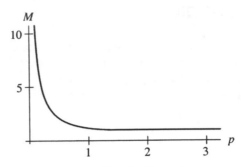

(b) If q is fixed and $p \to 0$, we need only consider the formula

$$M(p, q) = \frac{1}{\sqrt{p^2q - p^4/4}}.$$

For small $p > 0$,

$$M(p, q) \approx \frac{1}{\sqrt{p^2q}} = \frac{1}{\sqrt{q}} \frac{1}{p}.$$

9. We guess a solution of the form $y(t) = \alpha \sin \omega t + \beta \cos \omega t$. Substituting $y(t)$ into the left-hand side of the equation, we obtain

$$-\omega^2\alpha \sin \omega t - \omega^2\beta \cos \omega t + b(\omega\alpha \cos \omega t - \omega\beta \sin \omega t) + k(\alpha \sin \omega t + \beta \cos \omega t).$$

Our guess is a solution if this expression agrees with the right-hand side of the equation, namely

$$kA \cos \omega t - b\omega A \sin \omega t.$$

Equating the coefficients of sine and cosine from both sides of the equation, we obtain

$$-\omega^2\beta + b\omega\alpha + k\beta = kA$$
$$-\omega^2\alpha - b\omega\beta + k\alpha = -b\omega A,$$

which can be rewritten as

$$b\omega\alpha + (k - \omega^2)\beta = kA$$
$$(k - \omega^2)\alpha - b\omega\beta = -b\omega A.$$

Solving this system of two equations for α and β yields

$$\alpha = A\frac{b\omega^3}{(k - \omega^2)^2 + b^2\omega^2} \quad \text{and} \quad \beta = A\frac{b\omega^3}{(k - \omega^2)^2 + b^2\omega^2}.$$

So the particular solution is

$$y(t) = A\frac{b\omega^3}{(k - \omega^2)^2 + b^2\omega^2} \sin \omega t + A\frac{b\omega^3}{(k - \omega^2)^2 + b^2\omega^2} \cos \omega t.$$

11. If ω is large, the coefficients of both terms in the particular solution are small because they both have ω^4 in the denominator and at most ω^3 in the numerator. Hence, the oscillations are small. The weak spring and dash pot do not transmit the forcing to the mass fast enough and the pushes and pulls "average" out before they can effect the motion of the mass.

EXERCISES FOR SECTION 4.5

1. **(a)** The stiffness of the roadbed is measured by the coefficient of y. Increasing the stiffness corresponds to increasing β.

   (b) As β increases, the term corresponding to the stretch in cable becomes less important. Therefore, the system behaves more like a linear system.

3. **(a)** As the strength of the cable is increased, the spring constant γ for the cables (which appears in the definition of $c(y)$) increases.

   (b) Increasing γ in the definition of $c(y)$ has several effects. First, it makes the jump between the behavior of the system for $y < 0$ and $y > 0$ (that is, between taut and loose cables) more pronounced. This effect increases the system's "nonlinearity." Also, the equilibrium point moves closer to $y = 0$. On the other hand, larger γ makes it harder to displace the bridge from its rest position, so in most situations, we only see small amplitude oscillations.

5. The buoyancy force is proportional to the volume of the water displaced. Suppose p is that proportionality constant and a is the length of one side of a cube. Let y be the distance between the bottom of the cube and the surface of the water. Since the cube always stays in contact with the water and is never completely submerged, the buoyancy force is $-pa^2y$, and the force equation $F = ma$ is

$$mg - pa^2y - \epsilon\frac{dy}{dt} = m\frac{d^2y}{dt^2},$$

where the term $\epsilon(dy/dt)$ measures the damping. This equation can be rewritten as

$$m\frac{d^2y}{dt^2} + \epsilon\frac{dy}{dt} + pa^2y = mg.$$

7. If the cube rises completely out of the water, the buoyancy force vanishes. If we represent the buoyancy force by the function $b(y)$, we have

$$b(y) = \begin{cases} -pa^2y & \text{if } y > 0; \\ 0 & \text{if } y < 0. \end{cases}$$

Therefore, the differential equation is

$$m\frac{d^2y}{dt^2} + \epsilon\frac{dy}{dt} = mg + b(y).$$

CHAPTER 5

Nonlinear Systems

EXERCISES FOR SECTION 5.1

1. The linearizations of systems (i) and (iii) are both

$$\frac{dx}{dt} = 2x + y$$
$$\frac{dy}{dt} = -y,$$

$$\begin{pmatrix} x_\phi & y_a \\ x_z & y_z \end{pmatrix}$$

$$\begin{pmatrix} 2 & 1 \\ 2x & -1 \end{pmatrix}$$

so these two systems have the same "local picture" near $(0,0)$. This system has eigenvalues 2 and -1; hence, $(0,0)$ is a saddle for these systems. System (ii) has linearization

$$\frac{dx}{dt} = 2x + y$$
$$\frac{dy}{dt} = y,$$

which has eigenvalues 2 and 1, hence, $(0,0)$ is a source for this system.

3.   (a) The linearized system is

$$\frac{dx}{dt} = -2x + y$$
$$\frac{dy}{dt} = -y.$$

We can see this either by "dropping higher-order terms" or by computing the Jacobian matrix

$$\begin{pmatrix} -2 & 1 \\ 2x & -1 \end{pmatrix}$$

and evaluating it at $(0,0)$.

(b) The eigenvalues of the linearized system are -2 and -1, so $(0,0)$ is a sink.

(c) The vector $(1,0)$ is an eigenvector for eigenvalue -2 and $(1,1)$ is an eigenvector for the eigenvalue -1.

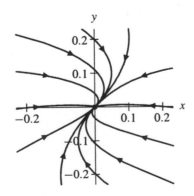

(d) By computing the Jacobian matrix

$$\begin{pmatrix} -2 & 1 \\ 2x & -1 \end{pmatrix}$$

and evaluating at $(2, 4)$, we see that linearized system at $(2, 4)$ is

$$\frac{dx}{dt} = -2x + y$$

$$\frac{dy}{dt} = 4x - y.$$

Its eigenvalues are $(-3 \pm \sqrt{17})/2$, so $(2, 4)$ is a saddle.

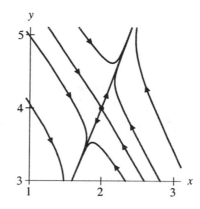

5. **(a)** Using separation of variables (or simple guessing), we have $x(t) = x_0 e^{-t}$.
    (b) The equation

$$\frac{dy}{dt} = -4x^3 + y$$

is a first-order, linear equation. We write the equation as

$$\frac{dy}{dt} - y = -4x^3.$$

Therefore, the integrating factor is e^{-t}. Multiplying both sides of the equation by e^{-t} yields

$$\left(\frac{dy}{dt} - y\right) e^{-t} = -4x^3 e^{-t}.$$

Note that the left-hand side is just the derivative of ye^{-t}, and the right-hand side is $-4x_0^3 e^{-3t} e^{-t}$, since $x(t) = x_0 e^{-t}$. Therefore we have

$$\frac{d}{dt} ye^{-t} = -4x_0^3 e^{-3t} e^{-t} = -4x_0^3 e^{-4t}.$$

After integrating and simplifying, we have

$$y(t) = x_0^3 e^{-3t} + (y_0 - x_0^3)e^t.$$

(c) The general solution of the system is

$$x(t) = x_0 e^{-t}$$
$$y(t) = x_0^3 e^{-3t} + (y_0 - x_0^3)e^t.$$

(d) For all solutions, $x(t) \to 0$ as $t \to \infty$. For a solution to tend to the origin as $t \to \infty$, we must have $y(t) \to 0$, and this can happen only if $y_0 - x_0^3 = 0$.

(e) Since $x = x_0 e^{-t}$, we see that a solution will tend toward the origin as $t \to -\infty$ only if $x_0 = 0$. In that case, $y(t) = y_0 e^t$, and $y(t) \to 0$ as $t \to -\infty$.

(f)

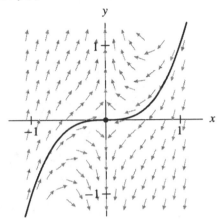

(g) Solutions tend away from the origin along the y-axis in both systems. In the nonlinear system, solutions approach the origin along the curve $y = x^3$ which is tangent to the x-axis. For the linearized system, solutions tend to the origin along the x-axis. Near the origin, the phase portraits are almost the same.

7. **(a)** The equilibrium points are $(0, 0)$, $(0, 100)$, $(150, 0)$, and $(30, 40)$. To determine the type of an equilibrium point, we compute the Jacobian matrix, which is

$$\begin{pmatrix} -2x - 3y + 150 & -3x \\ -2y & -2x - 2y + 100 \end{pmatrix},$$

and evaluate at the point. At $(0, 0)$, the Jacobian is

$$\begin{pmatrix} 150 & 0 \\ 0 & 100 \end{pmatrix},$$

and the eigenvalues are 150 and 100. Hence, the origin is a source. At $(0, 100)$, the Jacobian matrix is

$$\begin{pmatrix} -150 & 0 \\ -200 & -100 \end{pmatrix},$$

and the eigenvalues are -150 and -100. So $(0, 100)$ is a sink. The Jacobian at $(150, 0)$ is

$$\begin{pmatrix} -150 & -450 \\ 0 & -200 \end{pmatrix},$$

and the eigenvalues are -150 and -200. Therefore, $(150, 0)$ is a sink. Finally, the Jacobian matrix at $(30, 40)$ is

$$\begin{pmatrix} -30 & -90 \\ -80 & -40 \end{pmatrix},$$

and the eigenvalues are approximately -120 and 50. So $(30, 40)$ is a saddle.

(b)

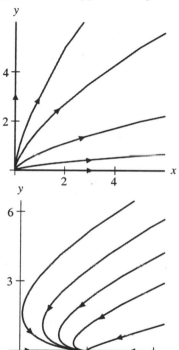
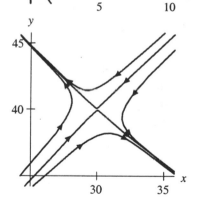

9. **(a)** The equilibrium points are $(0, 0)$, $(0, 25)$, $(100, 0)$ and $(75, 12.5)$. We classify these equilibrium points by computing the Jacobian matrix, which is

$$\begin{pmatrix} 100 - 2x - 2y & -2x \\ -y & 150 - x - 12y \end{pmatrix},$$

and evaluating it at each of the equilibrium points. At $(0, 0)$, the Jacobian matrix is

$$\begin{pmatrix} 100 & 0 \\ 0 & 150 \end{pmatrix},$$

and the eigenvalues are 100 and 150. So this point is a source. At $(0, 25)$, the Jacobian matrix is

$$\begin{pmatrix} 50 & 0 \\ -25 & -150 \end{pmatrix},$$

and the eigenvalues are 50 and -150. Hence, this point is a saddle. At $(100, 0)$, the Jacobian matrix is

$$\begin{pmatrix} -100 & -200 \\ 0 & 50 \end{pmatrix},$$

and the eigenvalues are -100 and 50. Therefore, this point is a saddle. Finally, at $(75, 12.5)$, the Jacobian matrix is

$$\begin{pmatrix} -75 & -150 \\ -12.5 & -75 \end{pmatrix},$$

and the eigenvalues are approximately -32 and -118. So this point is a sink.

(b)

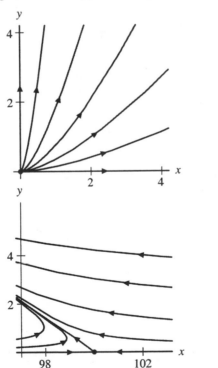

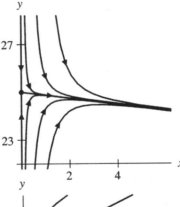

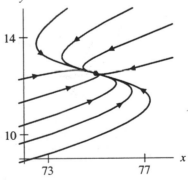

11. **(a)** The equilibrium points in the first quadrant are $(0, 0)$, $(0, 50)$ and $(40, 0)$. To classify these equilibrium points, we compute the Jacobian matrix, which is

$$\begin{pmatrix} -2x - y + 40 & -x \\ -2xy & -x^2 - 3y^2 + 2500 \end{pmatrix},$$

and we evaluate it at each of the points. At $(0, 0)$, the Jacobian matrix is

$$\begin{pmatrix} 40 & 0 \\ 0 & 2500 \end{pmatrix},$$

which has eigenvalues 40 and 2500. Therefore, $(0, 0)$ is a source. At $(0, 50)$, the Jacobian matrix is

$$\begin{pmatrix} -10 & 0 \\ 0 & -5000 \end{pmatrix},$$

which has eigenvalues -10 and -5000. So $(0, 50)$ is a sink. At $(40, 0)$, the Jacobian matrix is

$$\begin{pmatrix} -40 & -40 \\ 0 & 900 \end{pmatrix},$$

which has eigenvalues -40 and 900. Hence, $(40, 0)$ is a saddle.

(b)

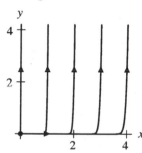 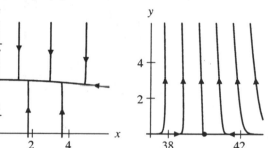

13. **(a)** The equilibrium points (in the first quadrant) are $(0,0)$, $(0, 50)$, $(60, 0)$, $(30, 40)$ and $(234/5, 88/5)$. To classify these points, we compute the Jacobian matrix, which is,

$$\begin{pmatrix} -16x - 6y + 480 & -6x \\ -2xy & -x^2 - 3y^2 + 2500 \end{pmatrix},$$

and evaluate it at each point. At $(0, 0)$, the Jacobian is

$$\begin{pmatrix} 480 & 0 \\ 0 & 2500 \end{pmatrix},$$

which has eigenvalues 480 and 2500. Thus, $(0, 0)$ is a source. At $(0, 50)$, the Jacobian matrix is

$$\begin{pmatrix} 180 & 0 \\ 0 & -5000 \end{pmatrix},$$

which has eigenvalues 180 and -5000. So $(0, 50)$ is a saddle. At $(60, 0)$, the Jacobian matrix is

$$\begin{pmatrix} -480 & -360 \\ 0 & -1100 \end{pmatrix},$$

which has eigenvalues -480 and -1100. Hence, $(60, 0)$ is a sink. At $(30, 40)$, the Jacobian matrix is

$$\begin{pmatrix} -240 & -180 \\ -2400 & -3200 \end{pmatrix},$$

which has eigenvalues approximately equal to -3339 and -101. So, $(30, 40)$ is a sink. At $(234/5, 88/5)$, the Jacobian matrix is

$$\begin{pmatrix} -1872/5 & -1404/5 \\ -41184/25 & -15488/25 \end{pmatrix},$$

which has eigenvalues approximately equal to -1188 and 194. Therefore, $(234/5, 88/5)$ is a saddle.

(b)

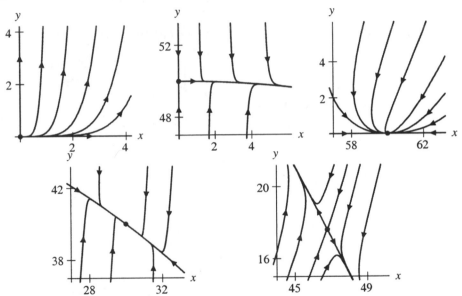

15. **(a)** The equilibrium points are $(0, 0)$, $(1, 1)$ and $(2, 0)$. We determine the type of each of these points by computing the Jacobian, which is

$$\begin{pmatrix} 2 - 2x - y & -x \\ -y & 2y - x \end{pmatrix},$$

and evaluating it at the points. At $(0, 0)$, the Jacobian is

$$\begin{pmatrix} 2 & 0 \\ 0 & 0 \end{pmatrix},$$

which has eigenvalues 2 and 0. An eigenvector for the eigenvalue 2 is $(1, 0)$, so solutions move away from the origin parallel to the x-axis. On the line $x = 0$, we have $dy/dt = y^2$ so solutions move upwards when $y \neq 0$. Hence, $(0, 0)$ is a node. However, solutions near the origin in the first quadrant move away from the origin as t increases. At $(1, 1)$, the Jacobian is

$$\begin{pmatrix} -1 & -1 \\ -1 & 1 \end{pmatrix},$$

which has eigenvalues $\pm\sqrt{2}$. So $(1, 1)$ is a saddle. At $(2, 0)$, the Jacobian is

$$\begin{pmatrix} -2 & -2 \\ 0 & -2 \end{pmatrix},$$

which has a double eigenvalue of -2. Therefore, $(2, 0)$ is a sink.

(b)

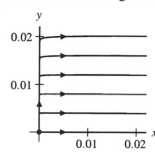
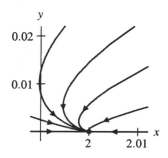

17. **(a)** The Jacobian matrix is

$$\begin{pmatrix} -3x^2 & 0 \\ 0 & -1+2y \end{pmatrix}$$

so the linearized system at $(0, 0)$ is

$$\frac{dx}{dt} = 0$$
$$\frac{dy}{dt} = -y.$$

(b) The eigenvalues are 0 and -1. Any vector on the x-axis is an eigenvector for the eigenvalue 0, and any vector on the y-axis is an eigenvector for the eigenvalue -1. Hence, the linearized system has a line of equilibrium points along the x-axis. Every other solution moves vertically toward the x-axis.

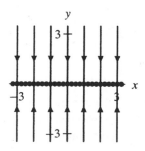

(c) The linearized system at $(0, 1)$ is

$$\frac{dx}{dt} = 0$$
$$\frac{dy}{dt} = y.$$

(d) The eigenvalues of this system are 0 and 1. Any vector on the x-axis is an eigenvector for the eigenvalue 0, and any vector on the y-axis is an eigenvector for the eigenvalue 1. Hence, the linearized system has a line of equilibrium points along the x-axis. Every other solution moves vertically away from the x-axis. It is important to remember that the origin for the linearized system corresponds to the equilibrium point $(0, 1)$ for the nonlinear system.

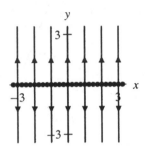

(e) The phase portrait is essentially a "combination" of two phase lines. The x-phase line has a sink at the origin. The y-phase line has a sink at the origin and a source at $y = 1$. Hence, the full phase portrait has a sink at $(0, 0)$, and the equilibrium point at $(0, 1)$ looks like a saddle.

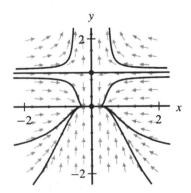

(f) The reason the linearizations and the nonlinear system look so different is that the equation for dx/dt contains only higher-order terms (just x^3 in this case). Since the equilibrium points occur along the y-axis ($x = 0$), the linearization has an entire line of equilibria in the x-direction.

19.   (a)

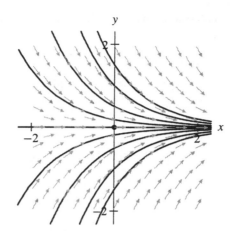

(b) The linearization of the equilibrium point at the origin has the coefficient matrix

$$\begin{pmatrix} 0 & 0 \\ 0 & -1 \end{pmatrix},$$

which has eigenvalues -1 and 0. So for the linearized system, the x-axis is a line of equilibria and solutions tend to zero in the y-direction. The nonlinear terms make solutions tend to zero in the x-direction for initial conditions with $x < 0$ and away from zero in the x-direction for initial conditions with $x > 0$.

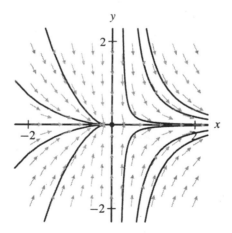

(c) The equilibria are $(\pm 1, 0)$. The coefficient matrix of the linearization at $(1, 0)$ is

$$\begin{pmatrix} 2 & 0 \\ 0 & -2 \end{pmatrix}.$$

The eigenvalues are 2 and -2, thus $(1, 0)$ is a saddle. The coefficient matrix of the linearization at $(-1, 0)$ is

$$\begin{pmatrix} -2 & 0 \\ 0 & -2 \end{pmatrix},$$

which has -2 as a repeated eigenvalue. So, $(-1, 0)$ is a sink.

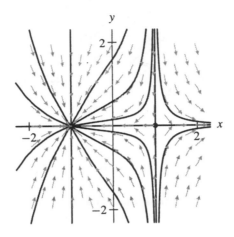

21. **(a)** The only equilibrium points occur if $a = 0$. Then all points on the curve $y = x^2$ are equilibrium points.

(b) The bifurcation occurs at $a = 0$.

(c) If $a < 0$, all solutions decrease in the y-direction since $dy/dt < 0$. If $a > 0$, all solutions increase in the y-direction since $dy/dt > 0$. If $a = 0$, there is a curve of equilibrium points located along $y = x^2$, and all solutions move horizontally.

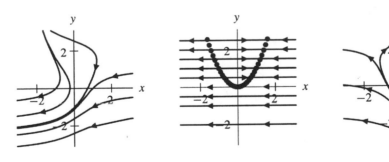

Phase portrait for $a < 0$ Phase portrait for $a = 0$ Phase portrait for $a > 0$

23. **(a)** The equilibrium points are $(0, 0)$, $(\pm 1/\sqrt{a}, \pm 1/\sqrt{a})$, so there is only one equilibrium point if $a \leq 0$, and three equilibrium points if $a > 0$.

(b) A bifurcation occurs at $a = 0$.

(c) If $a < 0$, there are is only one equilibrium point at the origin, and this equilibrium point is a spiral source. If $a > 0$, the system has two additional equilibrium points, at $(\pm 1/\sqrt{a}, \pm 1/\sqrt{a})$. These equilibrium points come from infinity as a increases through 0.

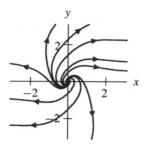

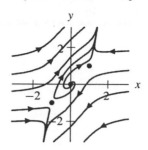

Phase portrait for $a < 0$ Phase portrait for $a = 0$ Phase portrait for $a > 0$

25. **(a)** The equilibrium points are $(\pm\sqrt{a/2}, -a/2)$, so there are no equilibrium points if $a < 0$, one equilibrium if $a = 0$, and two equilibrium points if $a > 0$

(b) A bifurcation occurs at $a = 0$.

(c) If $a < 0$, there are no equilibrium points and all solutions come from and go to infinity. If $a = 0$, an equilibrium point appears at the origin. This equilibrium point has eigenvalues 0 and 1 and is a node. If $a > 0$, the system has two equilibrium points, at $(\pm\sqrt{a/2}, -a/2)$.

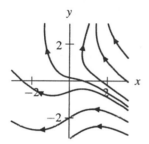

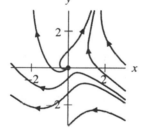

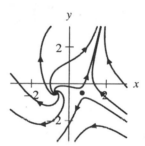

Phase portrait for $a < 0$ Phase portrait for $a = 0$ Phase portrait for $a > 0$

27. **(a)** The fact that $(0, 0)$ is an equilibrium point says that, if both X and Y are absent from the island, then neither will ever migrate to the island. However, it may be possible for one species to migrate if the other is already on the island.

(b) If a small population consisting solely of one of the species reproduces rapidly, then we expect both $\partial f/\partial x$ and $\partial g/\partial y$ to be positive and large at $(0, 0)$. We expect this because these partials are the coefficients of x and y in the linearization at $(0, 0)$.

(c) Since the species compete, an increase in y decreases dx/dt and an increase in x decreases dy/dt. Hence, both $\partial f/\partial y$ and $\partial g/\partial x$ are negative at $(0, 0)$ since $\partial f/\partial y$ is the coefficient of y in the dx/dt equation and $\partial g/\partial x$ is the coefficient for x in the dy/dt equation for the linearization at the origin.

(d) Suppose the coefficient matrix of the linearized system is

$$\begin{pmatrix} a & b \\ c & d \end{pmatrix},$$

with a and d positive and large and b and c negative. The eigenvalues are

$$\frac{(a+d) \pm \sqrt{(a-d)^2 + 4bc}}{2}.$$

If b and c are near zero, then $(0, 0)$ is a source. If b and c are very negative, then $(0, 0)$ is a saddle.

It is also possible to have 0 as an eigenvalue of the linearized system in which case the linearization fails to determine the behavior of the nonlinear system near $(0, 0)$.

(e) For the linearized system, note that $dx/dt < 0$ along the positive y-axis and $dy/dt < 0$ along the positive x-axis. If the origin is a saddle, the eigenvectors for the negative eigenvalue must be in the first and third quadrants, and a typical solution near the origin starting in the first quadrant has one of the species going extinct. If the origin is a source, then a typical solution near the origin has one or the other of the species going extinct except for one curve of solutions in the first quadrant.

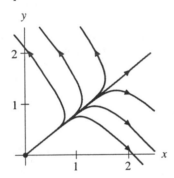 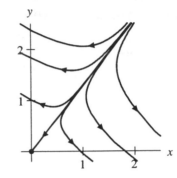

29. (a) At $(0, 0)$, $\partial f/\partial x$ is positive and large, and $\partial g/\partial y$ is positive and small.

(b) At $(0, 0)$, $\partial f/\partial y$ is negative with a large absolute value and $\partial g/\partial x = 0$.

(c) With these assumptions, the Jacobian matrix is

$$\begin{pmatrix} a & b \\ 0 & d \end{pmatrix},$$

where $a > 0$, $b < 0$, and $d > 0$ is much smaller than a. The eigenvalues of this matrix are a and d, so $(0, 0)$ is a source.

(d) Note that for $y = 0$, $dy/dt = 0$, and the eigenvector for a is in the x-direction.

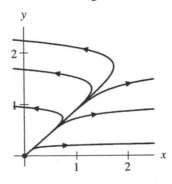

EXERCISES FOR SECTION 5.2

1. For x- and y-nullclines, $dx/dt = 0$, and $dy/dt = 0$ respectively. Then, we obtain $y = -x + 2$ for the x-nullcline and $y = x^2$ for the y-nullcline. To find intersections, we set $-x + 2 = x^2$, or $(x + 2)(x - 1) = 0$. Solving this for x yields $x = 1, -2$. For $x = 1$, $y = 1$, and for $x = -2$, $y = 4$. So the equilibrium points are $(1, 1)$ and $(-2, 4)$.

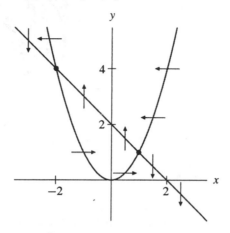

The solution for (a) is in the left-down region, and therefore, it eventually enters the region where $y < -x + 2$ and $y < x^2$. Once the solution enters this region, it stays there because the vector field on the boundaries never points out. Solutions for (b) and (c) start in this same region. Hence, all three solutions will go down and to the right without bound.

3. For the x-nullcline, $x(x - 1) = 0$, or $x = 0$ and $x = 1$, and for the y-nullcline, $y = x^2$. The equilibrium points are the intersection points of the x- and y-nullclines. They are $(0, 0)$ and $(1, 1)$.

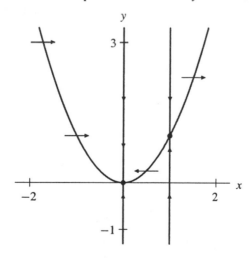

The initial conditions (a) and (b) are in right-up and left-up region respectively. Therefore, their solution curves eventually enter the region where $y > x^2$ and $x \leq 1$, and tend toward the equilibrium point at $(0, 0)$. The initial condition (c) is on the x-nullcline $x = 1$, and therefore its solution curve tends to the equilibrium point at $(1, 1)$.

5. (a) The x-nullcline is made up of the lines $x = 0$ and $y = -x/3 + 50$. The y-nullcline is made up of the lines $y = 0$ and $y = -2x + 100$.

(b)

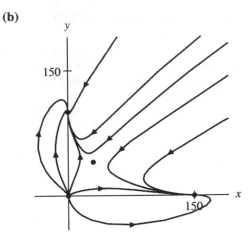

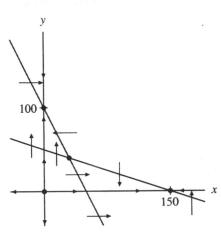

(c) Most solutions tend toward one of the equilibrium points $(0, 100)$ or $(150, 0)$. One curve of solutions divides these two behaviors. On this curve, solutions tend toward the saddle equilibrium at $(30, 40)$.

7. (a) The x-nullcline consists of the two lines $x = 0$ and $y = -x/2 + 50$. The y-nullcline consists of the two lines $y = 0$ and $y = -x/6 + 25$.

(b)

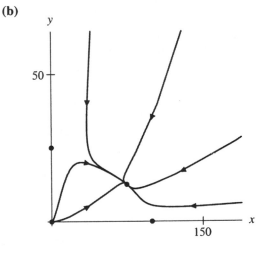

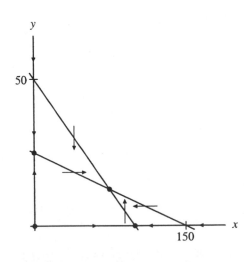

(c) All solutions off the axes tend toward the sink at $(75, 25/2)$. On the x-axis, solutions tend to the saddle at $(100, 0)$. On the y-axis, solutions tend to the saddle at $(0, 25)$.

9. (a) The x-nullcline is given by the two lines $x = 0$ and $y = -x + 40$. The y-nullcline is given by the line $y = 0$ and the circle $x^2 + y^2 = 50^2$.

(b)

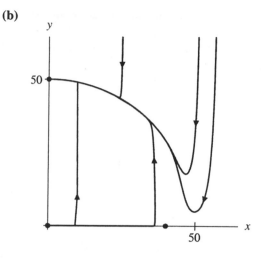

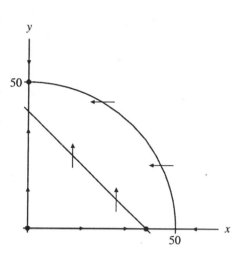

(c) Solutions off the x-axis tend toward the sink at $(0, 50)$. Solutions on the x-axis tend toward the saddle at $(40, 0)$.

11. (a) The x-nullcline is given by the line $x = 0$ and the line $y = -4x/3 + 80$. The y-nullcline is given by the line $y = 0$ and the circle $x^2 + y^2 = 50^2$, (recall we are only interested in the first quadrant).

(b)

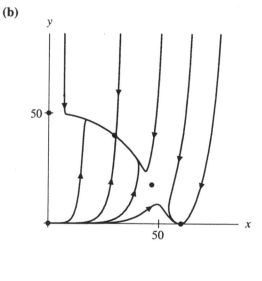

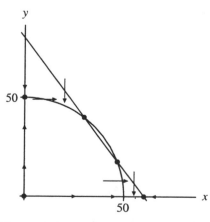

(c) Most solutions tend toward the sink at $(30, 40)$ or the sink at $(60, 0)$. The curve dividing these two behaviors is a curve of solutions tending toward the saddle at $(234/5, 88/5)$. Solutions on the y-axis tend toward the saddle at $(0, 50)$.

13. (a) The x-nullcline is given by the lines $x = 0$ and $y = -x + 2$. The y-nullcline is given by the lines $y = 0$ and $y = x$.

(b)

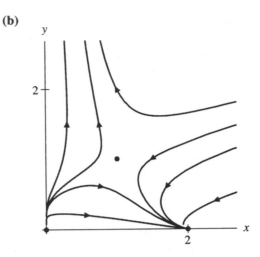

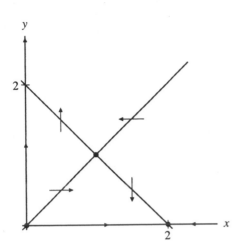

(c) Most solutions tend toward either the sink at $(2, 0)$ or toward infinity in the y-direction (with $x < 1$). The curve separating these two behaviors is a curve of solutions that tend toward the saddle at $(1, 1)$.

15. (a) Since the species are cooperative, an increase in y results in an increase in x and vice versa. Therefore, one needs to change the signs in front of B and D from $-$ to $+$.

(b) The x-nullcline is given by $x = 0$ or $-Ax + By + C = 0$. The y-nullcline is given by $y = 0$ or $Dx - Ey + F = 0$. The origin is always an equilibrium point. Also, $x = 0$, $y = F/E$ and $x = C/A$, $y = 0$ are equilibrium points. Equilibrium points with both x and y positive arise from solutions of

$$\begin{cases} -Ax + By + C = 0 \\ Dx - Ey + F = 0 \end{cases}$$

In matrix notation, we obtain

$$\begin{pmatrix} -A & B \\ D & -E \end{pmatrix} \begin{pmatrix} x \\ y \end{pmatrix} = \begin{pmatrix} -C \\ -F \end{pmatrix}.$$

In order for a unique solution to exist, $AE - BD \neq 0$. Then, the solution is

$$\begin{pmatrix} x \\ y \end{pmatrix} = \frac{1}{AE - BD} \begin{pmatrix} CE + BF \\ CD + AF \end{pmatrix}$$

Since A through F are all positive, we must have $AE - BD > 0$ for the solution to be in the first quadrant.

If $AE - BD = 0$, then $-Ax + By$ must be a negative multiple of $Dx - Ey$, so there are no solutions with both x and y positive.

17. **(a)** For the a-nullcline, $da/dt = 0$, so $2 - ab/2 = 0$, or $ab = 4$. For the b-nullcline, $db/dt = 0$, so $ab = 3$. Both nullclines are hyperbolas, and the curve of $ab = 4$ is above the one of $ab = 3$. Therefore, the direction of vector field on $ab = 4$ is vertical and downward, and the one on $ab = 3$ is horizontal and to the right.

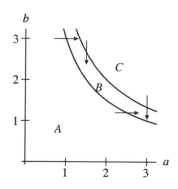

(b) Below and above $ab = 3$, $db/dt > 0$ and $db/dt < 0$ respectively. Below and above $ab = 3$, $da/dt > 0$ and $da/dt < 0$ respectively. Therefore, in region A, the vector field points up and to the right, in region B, the vector field points down and to the right, and in region C, the vector field points down and to the left.

(c) On the boundaries of B, the direction of the vector field never points out of B. Therefore, as time increases, these solutions are asymptotic to the positive x-axis from above.

19. **(a)** For the a-nullcline, $da/dt = 0$, so $b^2/3 - ab/2 + 2 = 0$. For the b-nullcline, $db/dt = 0$, so $-b^2/3 - ab/2 + 3/2 = 0$.

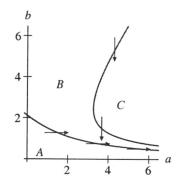

The a-nullcline is above the b-nullcline. Below the a-nullcline, $da/dt > 0$ and above the b-nullcline, $db/dt < 0$.

(b) Note that $da/dt > 0$ and $da/dt < 0$ below and above the a-nullcline respectively. Similarly, $db/dt > 0$ and $db/dt < 0$ below and above the b-nullcline respectively. Therefore, in region A, the vector field points up and to the right, in region B, the vector field points down and to the right, and in region C, the vector field points down and to the left.

(c) On the boundaries of B, the direction of the vector field never points out of B. Therefore, as time increases, these solutions are asymptotic to the positive x-axis from above.

21. For the x-nullcline, $dx/dt = 0$; thus, $y = 0$. For the y-nullcline, $dy/dt = 0$; thus, $x(1 - x) = 0$. The line $y = 0$ is x-nullcline, and the lines $x = 0$ and $x = 1$ are y-nullclines. Since $dx/dt = y$, x increases for $y > 0$ and decreases for $y < 0$. Similarly, $dy/dt > 0$ for $0 < x < 1$ and $dy/dt < 0$ for $x < 0$ and $x > 1$. So, the function y increases for $0 < x < 1$ and decreases for $x < 0$ and $x > 1$.

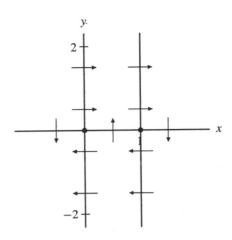

23. The Jacobian matrix of the vector field is

$$\begin{pmatrix} 0 & 1 \\ 1 - 2x & 0 \end{pmatrix}.$$

The coefficient matrix of the linearization at $(0, 0)$ is

$$\begin{pmatrix} 0 & 1 \\ 1 & 0 \end{pmatrix}$$

which has eigenvalues ± 1 and hence is a saddle with eigenvectors $(1, 1)$ (for 1) and $(1, -1)$ (for -1). The coefficient matrix of the linearization at $(1, 0)$ is

$$\begin{pmatrix} 0 & 1 \\ -1 & 0 \end{pmatrix}$$

which has eigenvalues $\pm i$ and is hence a center. From this information we can conclude that solutions with initial conditions near $(0, 0)$ move away from the origin in the $(1, 1)$ or $(-1, -1)$ direction while solutions near $(1, 0)$ tend to rotate around $(1, 0)$.

What we can not tell is what solutions do over the long-term. For example, do solutions spiral toward or away from $(1, 0)$? The higher order terms could cause either behavior (or all solutions could be periodic). Also, what is the behavior of the stable and unstable separatrices associated with the saddle? It turns out that this information is available for this system using techniques of Section 5.3

EXERCISES FOR SECTION 5.3

1. **(a)** We compute that

$$\frac{\partial H}{\partial x} = x - x^3$$

and so

$$\frac{dy}{dt} = -\frac{\partial H}{\partial x}.$$

Also,

$$\frac{\partial H}{\partial y} = y = \frac{dx}{dt}.$$

Hence, this is a Hamiltonian system with Hamiltonian function H.

(b) Note that $(0, 0)$ is a local minimum and $(\pm 1, 0)$ are saddle points.

(c) The equilibrium point $(0, 0)$ is a center and $(\pm 1, 0)$ are saddles. The saddles are connected by separatrix solutions.

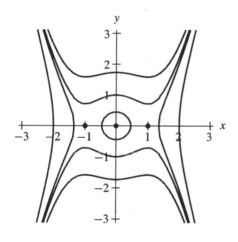

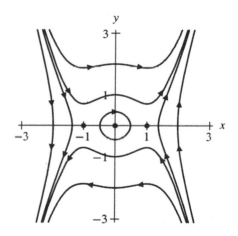

3. **(a)** If $H(x, y) = x \cos y + y^2$, then

$$\frac{\partial H}{\partial x} = \cos y$$

and so

$$\frac{dy}{dt} = -\frac{\partial H}{\partial x}.$$

Similarly,

$$\frac{\partial H}{\partial y} = -x \sin y + 2y = \frac{dx}{dt}.$$

(b)

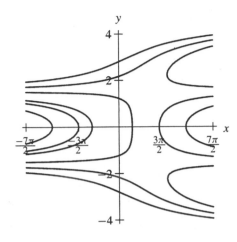

(c) The equilibrium points occur at points of the form $((1 - 4n)\pi, (2n - \frac{1}{2})\pi)$ and $((1 + 4n)\pi, (2n + \frac{1}{2})\pi)$ where n is an integer.

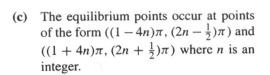

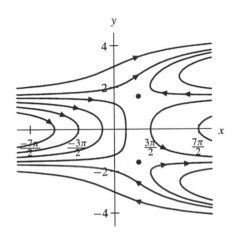

5. A large amplitude swing will take θ near $\pm\pi$, $v = 0$, the equilibrium point corresponding to the pendulum being balanced straight up. Near equilibrium points the vector field is very short, so solutions move very slowly. A solution passing close to $\pm\pi$, $v = 0$ must move slowly and hence, take a long time to make one complete swing. Hence, very high swings have long period. We must also be careful not to let the pendulum "swing over the top".

7. **(a)** The linearization at the origin is

$$\frac{d\theta}{dt} = v$$

$$\frac{dv}{dt} = -\frac{g}{l}\theta.$$

The eigenvalues of this system are $\pm i\sqrt{g/l}$, so the natural period is $2\pi/(\sqrt{g/l})$, which can also be written as $2\pi\sqrt{l}/\sqrt{g}$. Doubling the arm length corresponds to replacing l with $2l$, but the computations above stay the same. The natural period for arm length $2l$ is $2\pi\sqrt{2l}/\sqrt{g}$. Doubling the arm length multiplies the natural period by $\sqrt{2}$.

(b) Compute

$$\frac{d(2\pi\sqrt{l}/\sqrt{g})}{dl} = \frac{\pi}{\sqrt{gl}}.$$

9. We know that the equilibrium points of a Hamiltonian system cannot be sources or sinks. Phase portrait (b) has a spiral source, so it is not Hamiltonian. Phase portrait (c) has a sink and a source, so it is not Hamiltonian. Phase portraits (a) and (d) might come from Hamiltonian systems. (Try to imagine a function which has the solution curves as level sets.)

11. First note that

$$\frac{\partial(x - 3y^2)}{\partial x} = 1 = -\frac{\partial(-y)}{\partial y}.$$

Hence, the system is Hamiltonian. Integrating dx/dt with respect to y yields

$$H(x, y) = xy - y^3 + c(x).$$

If we differentiate $H(x, y)$ with respect to x, we get

$$y + c'(x),$$

which we want to be the negative of $dy/dt = -y$. Hence $c'(x) = 0$, and we pick the antiderivative $c(x) = 0$. A Hamiltonian function is

$$H(x, y) = xy - y^3.$$

13. First we check to see if the partial derivative with respect to x of the first component of the vector field is the negative of the partial derivative with respect to y of the second component. We have

$$\frac{\partial(x \cos y)}{\partial x} = \cos y$$

while

$$-\frac{\partial(-y \cos x)}{\partial y} = \cos x.$$

Since these two are not equal, the system is not Hamiltonian.

15. **(a)** We note that

$$\frac{\partial(1 - y^2)}{\partial x} = 0 \neq -\frac{\partial(x(1 + y^2))}{\partial y} = -2xy.$$

Hence, the system is not Hamiltonian.

(b)

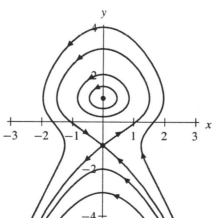

This phase portrait is consistent with the existence of a conserved quantity, but it is difficult to produce such a quantity from the figure alone.

(c) We note that

$$\frac{\partial((1-y^2)/(1+y^2))}{\partial x} = 0 = \frac{\partial(x)}{\partial y},$$

so the system is Hamiltonian. The Hamiltonian function is

$$H(x, y) = \int \frac{1-y^2}{1+y^2}\, dy - \frac{x^2}{2},$$

which is equal to

$$H(x, y) = -y + 2\arctan y - \frac{x^2}{2}.$$

(d) Multiplying the vector field by the function $f(x, y)$ changes the lengths of the vectors in the vector field, but it does not change their directions. Hence, the direction fields for the two systems agree. Consequently, the phase portraits are the same for the two systems, but solutions move at different speeds along the solution curves in the phase plane. Since H is a conserved quantity for the Hamiltonian system, the solution curves lie on level sets of H, and H is also a conserved quantity for the original system.

We can check that H is a conserved quantity by hand by computing the derivative with respect to t of $H(x(t), y(t))$, where $(x(t), y(t))$ is a solution of the original system. From the Chain Rule, we have

$$\frac{dH(x(t), y(t))}{dt} = \frac{\partial H}{\partial x}\frac{dx}{dt} + \frac{\partial H}{\partial y}\frac{dy}{dt}$$

$$= -x(1-y^2) + \frac{1-y^2}{1+y^2}(x(1+y^2))$$

$$= 0.$$

(e) See the answer to part (d).

17. Using the technique of Exercise 15, we we multiply the vector field by $1/(2-y)$. As in Exercise 14, the resulting system

$$\frac{dx}{dt} = \frac{1-y^2}{2-y}$$

$$\frac{dy}{dt} = x$$

is Hamiltonian. The Hamiltonian is

$$H(x, y) = -\frac{x^2}{2} + \int \frac{y^2-1}{y-2}\, dy$$

$$= -\frac{x^2}{2} + \int 2 + y + \frac{3}{y-2}\, dy$$

$$= -\frac{x^2}{2} + 2y + \frac{y^2}{2} + 3\ln|y-2|.$$

The function

$$H(x, y) = -\frac{x^2}{2} + 2y + \frac{y^2}{2} + 3\ln|y - 2|$$

is a conserved quantity for the original system. However, it is not defined on the line $y = 2$. From the system, we see that this line is a single solution curve that separates the two half-planes, $y < 2$ and $y > 2$.

19. First note that this system is Hamiltonian for every value of a. The Hamiltonian function depends on a and is given by

$$H(x, y) = x^2 y + xy^2 - ax.$$

If $a > 0$, then the system has two saddle equilibrium points on the y-axis at $(0, \pm\sqrt{a})$. If $a = 0$, then system has only one equilibrium point at $(0, 0)$. If $a < 0$, the system again has two saddles, but they are now located at $(\pm 2\sqrt{-3a}/3, \mp\sqrt{-3a}/3)$. This corresponds to a change in shape of the graph of H.

EXERCISES FOR SECTION 5.4

1. (a) Let $(x(t), y(t))$ be a solution of the system. Then

$$\frac{d}{dt}(L(x(t), y(t))) = \frac{\partial L}{\partial x}\frac{dx}{dt} + \frac{\partial L}{\partial y}\frac{dy}{dt}$$

$$= x(-x^3) + y(-y^3)$$

$$= -x^4 - y^4.$$

This is negative except at $x = y = 0$, which is an equilibrium point of the system.

(b) The level sets are circles centered at the origin.

(c) Since L decreases along solutions, every solution must approach the origin as t increases.

3. (a) We compute the eigenvalues. The characteristic polynomial is $\lambda^2 + 0.1\lambda + 4$, so the eigenvalues are

$$\lambda = \frac{-0.1 \pm i\sqrt{15.99}}{2}.$$

Hence the origin is a spiral sink.

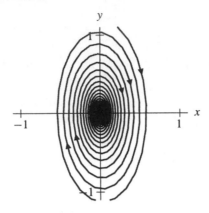

(b) Let $(x(t), y(t))$ be a solution. Then

$$\frac{d}{dt}(L(x(t), y(t))) = 2x(y) + 2y(-4x - 0.1y)$$

$$= -6xy - 0.2y^2.$$

In order for L to be a Lyapunov function, this quantity must be less than or equal to zero for all (x, y). However, if we take the solution with $x(0) = -1, y(0) = 1$ then

$$\frac{d}{dt}(L(x(t), y(t)))|_{t=0} = 5.8$$

which is positive, so L is not a Lyapunov function.

(c) Again let $(x(t), y(t))$ be a solution and compute

$$\frac{d}{dt}(K(x(t), y(t))) = 4x(y) + y(-4x - 0.1y)$$

$$= -0.1y^2$$

and this quantity is negative except at $y = 0$. (Solutions cross the $y = 0$ line at a discrete set of times. This is a technical point needed for the definition of Lyapunov functions.) So K is a Lyapunov function for the system.

5. We assume that the swings are small, hence we consider solutions near $\theta = v = 0$. The linearized system at the origin has Jacobian

$$\begin{pmatrix} 0 & 1 \\ -g/l & -b/m \end{pmatrix}$$

which has eigenvalues

$$\frac{-b/m \pm \sqrt{(b/m)^2 - 4g/l}}{2}.$$

When b/m is small, the eigenvalues are complex and the natural period of the oscillations is

$$2\pi / \sqrt{\frac{g}{l} - \frac{b^2}{4m^2}}$$

or

$$2\pi \left((g/l) - b^2 m^{-2}/4\right)^{-1/2}$$

If we differentiate this expression with respect to m we obtain

$$\frac{d}{dm}\left(2\pi\left(\frac{g}{l} - \frac{b^2}{4}m^{-2}\right)^{-1/2}\right) = -\pi\left(\frac{g}{l} - \frac{b^2}{4}m^{-2}\right)^{-3/2}\left(-\frac{b^2}{4}\left(-2m^{-3}\right)\right)$$

which is less than zero. Hence, as the mass m increases, the natural period of the linearization near $(0, 0)$ decreases and the clock runs fast.

7. In order to be usable as a clock, the pendulum arm must swing back and forth. That is, the equilibrium point at the origin must be a spiral sink or center so that solutions oscillate. So we study the equilibrium point at the origin. The linearized system at the origin has Jacobian

$$\begin{pmatrix} 0 & 1 \\ -g/l & -b/m \end{pmatrix},$$

which has eigenvalues

$$\frac{-b/m \pm \sqrt{(b/m)^2 - 4g/l}}{2}.$$

To have oscillating solutions we must have complex eigenvalues, that is we must have

$$\frac{b^2}{m^2} - \frac{4g}{l} < 0.$$

This holds if

$$m > \frac{b\sqrt{l}}{2\sqrt{g}}.$$

(We could also have $m < -b\sqrt{l}/(2\sqrt{g})$, but negative mass is "unphysical".)

9. The linearized system at the origin has complex eigenvalues (when b/m is small) that have real part equal to $-b/(2m)$. So near $\theta = v = 0$ the amplitude of the oscillations decrease at the same rate as

$$e^{-tb/(2m)}.$$

In order for the clock to keep accurate time, we need θ and v to remain close to zero. An absolutely "worst case" for the size of θ is $|\theta| < \pi$. So we can get a rough approximation of how long it takes for the amplitude of an initial swing to decay to 0.1 by finding t such that

$$\pi e^{-tb/(2m)} = 0.1.$$

This occurs when

$$t = -\frac{2m}{b}\ln\left(\frac{0.1}{\pi}\right).$$

Pendulum clocks need to be wound because energy must be added to the system since it is dissipative.

11.   (a) The linearized system at the origin has Jacobian

$$\begin{pmatrix} 0 & 1 \\ -g/l & -b/m \end{pmatrix},$$

which has eigenvalues

$$\frac{-b/m \pm \sqrt{(b/m)^2 - 4g/l}}{2}.$$

If b/m is small and negative, these eigenvalues have a positive real part and the origin is a spiral source. The linearized system at $(\pi, 0)$ has Jacobian

$$\begin{pmatrix} 0 & 1 \\ g/l & -b/m \end{pmatrix},$$

which has eigenvalues

$$\frac{-b/m \pm \sqrt{(b/m)^2 + 4g/l}}{2}.$$

If b/m is small (positive or negative), the two eigenvalues differ in sign, so this equilibrium point is a saddle. The equilibrium points $(\pm 2n\pi, 0)$ are just translates of the origin while the equilibrium points $(\pm(2n + 1)\pi, 0)$ are translates of $(\pi, 0)$.

(b)

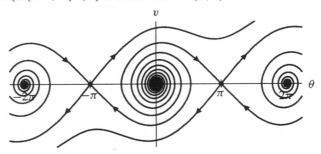

(c) A solution that has initial point near $(0, 0)$ will spiral away from the origin. Eventually, θ will become larger than π or smaller than $-\pi$ and then θ will increase or decrease monotonically toward infinity. This corresponds to the pendulum arm swinging with higher and higher amplitude until it "passes over the top" either clockwise or counter clockwise. After it passes over the top once, then it will continue to rotate in the same direction, accelerating with each rotation.

13.   (a) We have $\nabla G(x, y) = (2x, -2y)$, so

$$\frac{dx}{dt} = 2x$$
$$\frac{dy}{dt} = -2y.$$

(b) The system is linear and has eigenvalues 2 and -2. Hence the origin is a saddle.

(c) The graph of G is a saddle surface turning up in the x-direction and down in the y-direction.

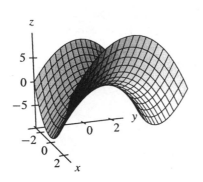 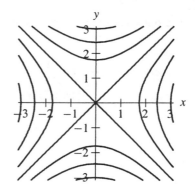

(d) The line of eigenvectors for eigenvalue 2 is the x-axis, the line of eigenvectors for eigenvalue -2 is the y-axis (see Chapter 3).

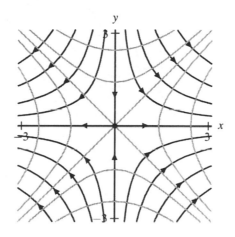

Phase portrait shown with level sets of G in gray.

15. **(a)** The Jacobian matrix is

$$\begin{pmatrix} 1 - 3x^2 & 0 \\ 0 & -1 \end{pmatrix}$$

At the origin the coefficient matrix of the linearization is

$$\begin{pmatrix} 1 & 0 \\ 0 & -1 \end{pmatrix}$$

which has eigenvalues 1 and -1. Hence, the origin is a saddle.

(b) At $(1, 0)$ the Jacobian matrix is

$$\begin{pmatrix} -2 & 0 \\ 0 & -1 \end{pmatrix}$$

which has eigenvalues -2 and -1. Hence, the point $(1, 0)$ is a sink.

(c) Since the eigenvalue -1 has eigenvectors along the y-axis, (and $-2 < -1$), solutions which approach $(1, 0)$ as t tends to infinity do so in the y direction (the only exception being solutions on the x-axis).

(d) The Jacobian matrix at $(-1, 0)$ is the same as at $(1, 0)$.

17. (a) The system is formed by taking the gradient of S. Hence, the system is

$$\frac{dx}{dt} = 2x - x^3 - 6xy^2$$

$$\frac{dy}{dt} = 2y - y^3 - 6x^2y.$$

(b)

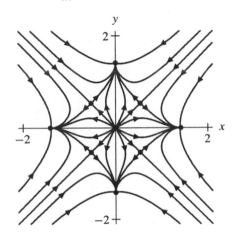

(c) From the phase portrait, we see that there are four sinks. Hence, there are four dead fish.

(d)

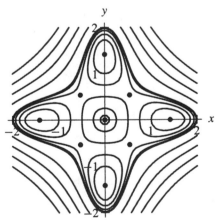

(e) For $|x|$ or $|y|$ large, S is negative, which is not physically reasonable.

19. (a) Since $f = \partial G/\partial x$ and $g = \partial G/\partial y$,

$$\frac{\partial f}{\partial y} = \frac{\partial^2 G}{\partial y \partial x} = \frac{\partial^2 G}{\partial x \partial y} = \frac{\partial g}{\partial x}.$$

(b) We compute

$$\frac{\partial f}{\partial y} = 3x$$

$$\frac{\partial g}{\partial x} = 2,$$

and these partials are not equal. By part (a), the system is not a gradient system.

21. Let $(y(t), v(t))$ be a solution of the "damped" system (with $k > 0$). Compute

$$\frac{d}{dt}(H(y(t), v(t))) = \frac{\partial H}{\partial y}\frac{dy}{dt} + \frac{\partial H}{\partial v}\frac{dv}{dt}$$

$$= \frac{dV}{dy}v + v\left(-\frac{dV}{dy} - kv\right)$$

$$= -kv^2.$$

Hence, for this system, H is a Lyapunov function.

23. Let $(x_1(t), x_2(t), p_1(t), p_2(t))$ be a solution and differentiate

$$H(x_1(t), x_2(t), p_1(t), p_2(t))$$

with respect to t using the Chain Rule. That is,

$$\frac{dH}{dt} = \frac{\partial H}{\partial x_1}\frac{dx_1}{dt} + \frac{\partial H}{\partial x_2}\frac{dx_2}{dt} + \frac{\partial H}{\partial p_1}\frac{dp_1}{dt} + \frac{\partial H}{\partial p_2}\frac{dp_2}{dt}.$$

Note that

$$\frac{\partial H}{\partial x_1} = k_1(x_1 - L_1) - k_2(x_2 - x_1 - L_2)$$

$$\frac{\partial H}{\partial x_2} = k_2(x_2 - x_1 - L_2)$$

$$\frac{\partial H}{\partial p_1} = \frac{p_1}{m_1}$$

$$\frac{\partial H}{\partial p_2} = \frac{p_2}{m_2}.$$

Using these partials along with the formulas for dx_1/dt, dx_2/dt, dp_1/dt, and dp_2/dt given by the system, we see that

$$\frac{dH}{dt} = \frac{\partial H}{\partial x_1}\frac{dx_1}{dt} + \frac{\partial H}{\partial x_2}\frac{dx_2}{dt} + \frac{\partial H}{\partial p_1}\frac{dp_1}{dt} + \frac{\partial H}{\partial p_2}\frac{dp_2}{dt} = 0.$$

EXERCISES FOR SECTION 5.5

1. **(a)** The Jacobian matrix for the system is

$$\begin{pmatrix} 1 - 2x - y & -x & 0 \\ y & 1 - 2y + x - z & -y \\ 0 & z & 1 - 2z + y \end{pmatrix}$$

so at $(1, 0, 0)$ the Jacobian matrix is

$$\begin{pmatrix} -1 & -1 & 0 \\ 0 & 2 & 0 \\ 0 & 0 & 1 \end{pmatrix}.$$

(b) The characteristic polynomial is $(1 - \lambda)(2 - \lambda)(-1 - \lambda)$, and the eigenvalues are $1, 2$ and -1. An eigenvector for eigenvalue 1 is $(0, 0, 1)$, an eigenvector for eigenvalue -1 is $(1, 0, 0)$ and an eigenvector for eigenvalue 2 is $(1, -3, 0)$.

(c) The equilibrium point is a saddle with two eigenvalues positive and one negative. Solutions tend toward the equilibrium point along the x-axis as t increases. Solutions in the plane formed by the two vectors $(0, 0, 1)$ and $(1, -3, 0)$ through the equilibrium point tend toward the equilibrium point as t decreases.

(d)

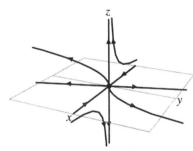

(e) If the initial condition satisfies $y = z = 0$ and x is close to 1, then the corresponding solution tends toward the equilibrium point as t increases. If y or z is small and positive and x is close to 1, then the solution may move closer to the equilibrium point if y and z are small, but eventually moves away from the equilibrium point.

3. **(a)** The Jacobian matrix at $(0, 0, 1)$ is

$$\begin{pmatrix} 1 & 0 & 0 \\ 0 & 0 & 0 \\ 0 & 1 & -1 \end{pmatrix}.$$

(b) The characteristic polynomial is $(1 - \lambda)\lambda(-1 - \lambda)$, so the eigenvalues are $-1, 0$ and 1. Corresponding eigenvectors are $(0, 0, 1)$, $(0, 1, 1)$ and $(1, 0, 0)$ respectively. (We can see this by computation, or use the fact that x decouples from y and z in the linearized system to reduce the dimension of the problem.)

(c) The linearized system has solutions that tend toward the original equilibrium point as t increases, solutions that tend toward the original equilibrium point as t decreases, and a line of equilibrium points through the original equilibrium point.

(d)

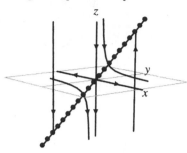

(e) For the linearized system, solutions with initial conditions with $x = 0$ tend in the z-direction toward an equilibrium point on the line of equilibrium points through the original equilibrium point. This leaves unclear the behavior of solutions for the nonlinear system on the $x = 0$ plane because nonlinear terms could cause solutions to tend toward or away from the equilibrium point. Off the plane $x = 0$, solutions move away from the equilibrium point in the x-direction.

5. **(a)** When $y = 0$ we have $dy/dt = 0$ and the equations for the rate of change of x and z are

$$\frac{dx}{dt} = x(1 - x)$$

$$\frac{dz}{dt} = z(1 - z).$$

These are both logistic equations with sinks at $x = 1$ and $z = 1$. Hence, $(1, 0, 1)$ is an equilibrium point. When $x = 0$ we have $dx/dt = 0$ and

$$\frac{dy}{dt} = y(1 - y) - yz$$

$$\frac{dz}{dt} = z(1 - z) + yz$$

and $y = z = 1$ is not an equilibrium point for the yz system so $(0, 1, 1)$ is not an equilibrium point for the three-dimensional system. The argument for $(1, 1, 0)$ is very similar.

(b) Having $y = 0$ corresponds to the moose being extinct. The food chain then has no connection between the wolves and the trees. Hence, both wolf and tree populations will tend toward their carrying capacity. When $x = 0$ the trees are extinct. The moose and wolfs then form a two dimensional predator-prey system. We do not expect that both moose and wolf populations to be maintained at their carrying capacitys because wolves eat moose. The argument when the wolves are extinct is similar.

7. **(a)** We must solve

$$x_0(1 - x_0) - x_0 y_0 = 0$$
$$y_0(1 - y_0) + x_0 y_0 - y_0 z_0 = 0$$
$$\zeta z_0(1 - z_0) + y_0 z_0 = 0.$$

Since we are looking for the solution where none of the coordinates is zero, we can divide the first equation by x_0, the second by y_0 and the third by z_0 to obtain

$$1 - x_0 - y_0 = 0$$
$$1 - y_0 + x_0 - z_0 = 0$$
$$\zeta - \zeta z_0 + y_0 = 0.$$

The first equation gives $y_0 = 1 - x_0$ which when plugged into the second equation gives $z_0 = 2x_0$ and these in the third equation give

$$x_0 = \frac{\zeta + 1}{2\zeta + 1}.$$

Hence, $y_0 = \zeta/(2\zeta + 1)$ and $z_0 = (2\zeta + 2)/(2\zeta + 1)$.

(b) We can compute the rate of change of the position of this equilibrium point as ζ changes as follows:

$$\frac{dx_0}{d\zeta} = \frac{-1}{(2\zeta + 1)^2}$$

$$\frac{dy_0}{d\zeta} = \frac{1}{(2\zeta + 1)^2}$$

$$\frac{dz_0}{d\zeta} = \frac{-2}{(2\zeta + 1)^2}.$$

Note that $dz_0/d\zeta$ is negative. Hence, a decrease in ζ, which we claimed corresponded to a decrease in the growth rate of the wolves, yields an increase in z_0. Similarly, a decrease in ζ gives a decrease in y_0 and an increase in x_0 because $dy_0/d\zeta$ is positive and $dx_0/d\zeta$ is negative (remember we are decreasing ζ).

(c) Note that the dz/dt equation is

$$\frac{dz}{dt} = \zeta z(1 - z) + yz = \zeta z - \zeta z^2 + yz$$

so ζ is multiplied onto the z and the z^2 terms. If we compare this to the modification of the equations given in the section we see that the change must be causes by the fact that ζ is multiplied onto the z^2 term. It does seem "unphysical" that a decrease in the growth-rate parameter could cause an increase in the equilibrium population, but the parameter ζ gives both the growth rate for small z (the ζz term) and the proportionality constant for the effect of large populations (the ζz^2) term. This observation is interesting and deserves further study.

EXERCISES FOR SECTION 5.6

1. Comparing the y values at the indicated points with the y values of the $y(t)$-graphs for $t = 2n\pi$ shows this return map corresponds to graph (ii). The solution oscillates with increasing and decreasing amplitude around $y = 1$.

3. Comparing the y values at the indicated points with the y values of the $y(t)$-graphs for $t = 2n\pi$ shows this return map corresponds to graph (i). The solution oscillates in a regular, but fairly complicated way.

5. (a) This return map corresponds to graph (iii).

   (b) The values of the θ-coordinate of the indicated points are decreasing monotonically and are all less than one. Hence, graph (iii) is the only one that fits.

   (c) The solution should oscillate with rising and falling amplitude which corresponds to the pendulum arm swinging higher and lower, but never swinging "over the top".

7. (a) This return map corresponds to graph (iv).

   (b) The θ-coordinates at the indicated points are bounded below by minus two, are all nonpositive and decrease, then increase. The graph with the corresponding behavior is graph (iv).

   (c) The pendulum swings back and forth, never "going over the top" with amplitude and period oscillating in a regular way.

9. (a) The second-order differential equation is

$$\frac{d^2y}{dt^2} + 3y = 0.2\sin t.$$

The general solution for the homogeneous equation is

$$y(t) = k_1 \sin\sqrt{3}\,t + k_2 \cos\sqrt{3}\,t.$$

Suppose $y(t) = a\sin t$. Differentiation yields

$$\frac{d^2y}{dt^2} + 3y = 2a\sin t = 0.2\sin t.$$

Then, $a = 1/10$ and the general solution is

$$y(t) = k_1 \sin\sqrt{3}\,t + k_2 \cos\sqrt{3}\,t + \tfrac{1}{10}\sin t.$$

By the initial condition, $y(0) = 1$ and $v(0) = 0$, one obtains

$$\begin{cases} k_2 = 1 \\ \sqrt{3}\,k_1 + \tfrac{1}{10} = 0 \end{cases}$$

or $k_1 = -\sqrt{3}/30$ and $k_2 = 1$. The solution is

$$y(t) = -\frac{\sqrt{3}}{30}\sin\sqrt{3}\,t + \cos\sqrt{3}\,t + \tfrac{1}{10}\sin t.$$

(b) By the solution,

$$y(t) = -\frac{\sqrt{3}}{30}\sin\sqrt{3}\,t + \cos\sqrt{3}\,t + \tfrac{1}{10}\sin t.$$

After 2π in time, $0.1\sin(t + 2\pi)$ gives the same value. Therefore, the Poincaré map is due to the first two terms and it moves on an ellipse.

(c) The change in the initial condition affects only constants, k_1 and k_2. Therefore, Poincaré map traces an ellipse and the scale of the ellipse depends on k_1 and k_2.

(d) The forced harmonic oscillator system is linear and the solution is the sum of the general solution for the homogeneous equation and the particular solution for the nonhomogeneous equation. In this case, the Poincaré map is due to the homogeneous solution as we saw in (c). On the other hand, the forced pendulum system is nonlinear and therefore, the solution becomes more complicated.

CHAPTER 6

Laplace Transforms

EXERCISES FOR SECTION 6.1

1. We have

$$\mathcal{L}[3] = \int_0^\infty 3e^{-st}\, dt$$

$$= \lim_{b \to \infty} \int_0^b 3e^{-st}\, dt$$

$$= \lim_{b \to \infty} \left(\frac{-3}{s} e^{-st} \Big|_0^b \right)$$

$$= \lim_{b \to \infty} -\frac{3}{s} \left(e^{-sb} - e^0 \right)$$

$$= \frac{3}{s} \quad \text{if } s > 0,$$

since $\lim_{b \to \infty} e^{-sb} = \lim_{b \to \infty} 1/e^{sb} = 0$ if $s > 0$.

3. We use the fact that $\mathcal{L}[df/dt] = s\mathcal{L}[f] - f(0)$. Letting $f(t) = t^2$ we have $f(0) = 0$ and

$$\mathcal{L}[2t] = s\mathcal{L}[t^2] - 0$$

or

$$2\mathcal{L}[t] = s\mathcal{L}[t^2]$$

using the fact that the Laplace transform is linear. Then since $\mathcal{L}[t] = 1/s^2$ (by the previous exercise), we have

$$\mathcal{L}[-5t^2] = -5\mathcal{L}[t^2] = -5\left(\frac{2\mathcal{L}[t]}{s} \right) = -\frac{10}{s^3}.$$

5. To show a rule by induction, we need two steps. First, we need to show the rule is true for $n = 1$. Then, we need to show that if the rule holds for n, then it holds for $n + 1$.

(a) $n = 1$. We need to show that $\mathcal{L}[t] = 1/s^2$.

$$\mathcal{L}[t] = \int_0^\infty te^{-st}\, dt.$$

Using integration by parts with $u = t$ and $dv = e^{-st}\, dt$, we find

$$\mathcal{L}[t] = \frac{te^{-st}}{-s} \Big|_0^\infty + \int_0^\infty \frac{e^{-st}}{s}\, dt$$

$$= \lim_{b \to \infty} \left[\frac{te^{-st}}{-s} \Big|_0^b \right] + \int_0^\infty \frac{e^{-st}}{s}\, dt$$

$$= \int_0^\infty \frac{e^{-st}}{s}\, dt$$

$$= -\frac{e^{-st}}{s^2} \Big|_0^\infty$$

$$= \frac{1}{s^2} \quad (s > 0).$$

(b) Now we assume that the rule holds for n, that is, that $\mathcal{L}[t^n] = n!/s^{n+1}$, and show it holds true for $n + 1$, that is, $\mathcal{L}[t^{n+1}] = (n + 1)!/s^{n+2}$. There are two different methods to do so:

(i)

$$\mathcal{L}[t^{n+1}] = \int_0^\infty t^{n+1} e^{-st}\, dt$$

Using integration by parts with $u = t^{n+1}$ and $dv = e^{-st}\, dt$, we find

$$\mathcal{L}[t^{n+1}] = -\frac{t^{n+1} e^{-st}}{s}\Big|_0^\infty + \int_0^\infty \frac{n+1}{s} t^n e^{-st}\, dt.$$

Now,

$$-\frac{t^{n+1} e^{-st}}{s}\Big|_0^\infty = \lim_{b \to \infty}\left[-\frac{t^{n+1} e^{-st}}{s}\Big|_0^b\right]$$

$$= \lim_{b \to \infty} \frac{-b^{n+1} e^{-sb}}{s} + 0$$

$$= 0 \quad (s > 0).$$

So

$$\mathcal{L}[t^{n+1}] = \int_0^\infty \frac{n+1}{s} t^n e^{-st}\, dt$$

$$= \frac{n+1}{s} \int_0^\infty t^n e^{-st}\, dt$$

$$= \frac{n+1}{s} \mathcal{L}[t^n].$$

Since we assumed that $\mathcal{L}[t^n] = n!/s^{n+1}$, we get that

$$\mathcal{L}[t^{n+1}] = \frac{n+1}{s} \cdot \frac{n!}{s^{n+1}} = \frac{(n+1)!}{s^{n+2}}$$

which is what we wanted to show.

(ii) We use the fact that $\mathcal{L}[df/dt] = s\mathcal{L}[f] - f(0)$. Letting $f(t) = t^{n+1}$ we have $f(0) = 0$ and

$$\mathcal{L}[(n+1)t^n] = s\mathcal{L}[t^{n+1}] - 0$$

or

$$(n+1)\mathcal{L}[t^n] = s\mathcal{L}[t^{n+1}]$$

using the fact that the Laplace transform is linear. Since we assumed $\mathcal{L}[t^n] = n!/s^{n+1}$, we have

$$\mathcal{L}[t^{n+1}] = \frac{n+1}{s}\mathcal{L}[t^n] = \frac{n+1}{s} \cdot \frac{n!}{s^{n+1}} = \frac{(n+1)!}{s^{n+2}},$$

which is what we wanted to show.

7. Since we know that

$$\mathcal{L}[e^{at}] = \frac{1}{s-a},$$

we see that

$$\mathcal{L}[e^{3t}] = \frac{1}{s-3},$$

giving

$$\mathcal{L}^{-1}\left[\frac{1}{s-3}\right] = e^{3t}.$$

9. We see that

$$\frac{2}{3s+5} = \frac{2}{3} \cdot \frac{1}{s+5/3},$$

so

$$\mathcal{L}^{-1}\left[\frac{2}{3s+5}\right] = \frac{2}{3}e^{-\frac{5}{3}t}.$$

11. We see that

$$\frac{5}{3s} = \frac{5}{3} \cdot \frac{1}{s},$$

so

$$\mathcal{L}^{-1}\left[\frac{5}{3s}\right] = \frac{5}{3},$$

since $\mathcal{L}^{-1}[1/s] = 1$.

13. Using the method of partial fractions, we have

$$\frac{2s+1}{(s-1)(s-2)} = \frac{A}{s-1} + \frac{B}{s-2}.$$

Putting the right-hand side over a common denominator gives $A(s-2) + B(s-1) = 2s+1$, which can be written as $(A+B)s + (-2A-B) = 2s+1$. So, $A+B = 2$, and $-2A-B = 1$. Thus $A = -3$ and $B = 5$, which gives

$$\mathcal{L}^{-1}\left[\frac{2s+1}{(s-1)(s-2)}\right] = \mathcal{L}^{-1}\left[\frac{5}{s-2} - \frac{3}{s-1}\right].$$

Finally,

$$\mathcal{L}^{-1}\left[\frac{2s+1}{(s-1)(s-2)}\right] = 5e^{2t} - 3e^t.$$

15.   (a) We have

$$\mathcal{L}\left[\frac{dy}{dt}\right] = s\mathcal{L}[y] - y(0)$$

and

$$\mathcal{L}[-y + e^{-2t}] = \mathcal{L}[-y] + \mathcal{L}[e^{-2t}] = -\mathcal{L}[y] + \frac{1}{s+2}$$

using linearity of the Laplace transform and the formula $\mathcal{L}[e^{at}] = 1/(s-a)$ from the text.

(b) Substituting the initial condition yields

$$s\mathcal{L}[y] - 2 = -\mathcal{L}[y] + \frac{1}{s+2}$$

so that

$$(s+1)\mathcal{L}[y] = 2 + \frac{1}{s+2}$$

which gives

$$\mathcal{L}[y] = \frac{1}{(s+1)(s+2)} + \frac{2}{s+1} = \frac{2s+5}{(s+1)(s+2)}.$$

(c) Using the method of partial fractions,

$$\frac{2s+5}{(s+1)(s+2)} = \frac{A}{s+1} + \frac{B}{s+2}.$$

Putting the right-hand side over a common denominator gives $A(s+2) + B(s+1) = 2s+5$, which can be written as $(A+B)s + (2A+B) = 2s+5$. So we have $A+B = 2$, and $2A+B = 5$. Thus, $A = 3$ and $B = -1$, and

$$\mathcal{L}[y] = \frac{3}{s+1} - \frac{1}{s+2}.$$

Therefore, $y(t) = 3e^{-t} - e^{-2t}$ is the desired function.

(d) Since $y(0) = 3e^0 - e^0 = 2$, $y(t)$ satisfies the given initial condition. Also,

$$\frac{dy}{dt} = -3e^{-t} + 2e^{-2t}$$

and

$$-y + e^{-2t} = -3e^{-t} + e^{-2t} + e^{-2t} = -3e^{-t} + 2e^{-2t},$$

so our solution also satisfies the differential equation.

17. **(a)** Taking Laplace transforms of both sides of the equation and simplifying gives

$$\mathcal{L}\left[\frac{dy}{dt}\right] + 7\mathcal{L}[y] = \mathcal{L}[1]$$

so

$$s\mathcal{L}[y] - y(0) + 7\mathcal{L}[y] = \frac{1}{s}$$

and $y(0) = 3$ gives

$$s\mathcal{L}[y] - 3 + 7\mathcal{L}[y] = \frac{1}{s}.$$

(b) Solving for $\mathcal{L}[y]$ gives

$$\mathcal{L}[y] = \frac{3}{s+7} + \frac{1}{s(s+7)} = \frac{3s+1}{s(s+7)}.$$

(c) Using the method of partial fractions, we get

$$\frac{3s+1}{s(s+7)} = \frac{A}{s} + \frac{B}{s+7}.$$

Putting the right-hand side over a common denominator gives $A(s+7) + Bs = 3s+1$, which can be written as $(A+B)s + 7A = 3s+1$. So $A + B = 3$, and $7A = 1$. Hence, $A = 1/7$ and $B = 20/7$, and we have

$$\mathcal{L}[y] = \frac{1/7}{s} + \frac{20/7}{s+7}.$$

Thus,

$$y(t) = \frac{20}{7}e^{-7t} + \frac{1}{7}.$$

(d) To check, we compute

$$\frac{dy}{dt} + 7y = -20e^{-7t} + 7\left(\frac{20}{7}e^{-7t} + \frac{1}{7}\right) = 1,$$

and $y(0) = 20/7 + 1/7 = 3$, so our solution satisfies the initial-value problem.

19. (a) Taking Laplace transforms of both sides of the equation and simplifying gives

$$\mathcal{L}\left[\frac{dy}{dt}\right] + 9\mathcal{L}[y] = \mathcal{L}[2]$$

so

$$s\mathcal{L}[y] - y(0) + 9\mathcal{L}[y] = \frac{2}{s}$$

and $y(0) = -2$ gives

$$s\mathcal{L}[y] + 2 + 9\mathcal{L}[y] = \frac{2}{s}.$$

(b) Solving for $\mathcal{L}[y]$ gives

$$\mathcal{L}[y] = -\frac{2}{s+9} + \frac{2}{s(s+9)} = \frac{-2s+2}{s(s+9)}.$$

(c) Using the method of partial fractions,

$$\frac{-2s+2}{s(s+9)} = \frac{A}{s} + \frac{B}{s+9}.$$

Putting the right-hand side over a common denominator gives $A(s+9) + Bs = -2s+2$, which can be written as $(A+B)s + 9A = -2s+2$. So $A + B = -2$ and $9A = 2$. Hence, $A = 2/9$ and $B = -20/9$, which gives us

$$\mathcal{L}[y] = \frac{2/9}{s} - \frac{20/9}{s+9}.$$

Finally,

$$y(t) = -\frac{20}{9}e^{-9t} + \frac{2}{9}.$$

(d) To check, we compute

$$\frac{dy}{dt} + 9y = 20e^{-9t} + 9\left(\frac{-20}{9}e^{-9t} + \frac{2}{9}\right) = 2,$$

and $y(0) = -20/9 + 2/9 = -2$, so our solution satisfies the initial-value problem.

21. **(a)** Putting the equation in the form

$$\frac{dy}{dt} + y = e^{-2t},$$

taking Laplace transforms of both sides of the equation and simplifying gives

$$\mathcal{L}\left[\frac{dy}{dt}\right] + \mathcal{L}[y] = \mathcal{L}[e^{-2t}]$$

so

$$s\mathcal{L}[y] - y(0) + \mathcal{L}[y] = \frac{1}{s+2}$$

and $y(0) = 1$ gives

$$s\mathcal{L}[y] - 1 + \mathcal{L}[y] = \frac{1}{s+2}.$$

(b) Solving for $\mathcal{L}[y]$ gives

$$\mathcal{L}[y] = \frac{1}{s+1} + \frac{1}{(s+1)(s+2)} = \frac{s+3}{(s+1)(s+2)}.$$

(c) Using partial fractions,

$$\frac{s+3}{(s+1)(s+2)} = \frac{A}{s+1} + \frac{B}{s+2}.$$

Putting the right-hand side over a common denominator gives $A(s+2) + B(s+1) = s+3$, which can be written as $(A+B)s + (2A+B) = s+3$. Thus, $A+B = 1$, and $2A+B = 3$. So $A = 2$ and $B = -1$, which gives us

$$\mathcal{L}[y] = \frac{2}{s+1} - \frac{1}{s+2}.$$

Hence,

$$y(t) = 2e^{-t} - e^{-2t}.$$

(d) To check, we compute

$$\frac{dy}{dt} + y = -2e^{-t} + 2e^{-2t} + \left(2e^{-t} - e^{-2t}\right) = e^{-2t},$$

and $y(0) = 2 - 1 = 1$, so our solution satisfies the initial-value problem.

23. **(a)** We have

$$\mathcal{L}\left[\frac{dy}{dt}\right] = s\mathcal{L}[y] - y(0)$$

and

$$\mathcal{L}[-y + t^2] = \mathcal{L}[-y] + \mathcal{L}[t^2] = -\mathcal{L}[y] + \frac{2}{s^3}$$

using linearity of the Laplace transform and the formula $\mathcal{L}[t^n] = n!/s^{n+1}$ from Exercise 5.

(b) Substituting the initial condition yields

$$s\mathcal{L}[y] - 1 = -\mathcal{L}[y] + \frac{2}{s^3}$$

so that

$$\mathcal{L}[y] = \frac{2/s^3 + 1}{1 + s} = \frac{2 + s^3}{s^3(s + 1)}.$$

(c) The best way to deal with this problem is with partial fractions. We seek constants A, B, C, and D such that

$$\frac{A}{s} + \frac{B}{s^2} + \frac{C}{s^3} + \frac{D}{s + 1} = \frac{2 + s^3}{s^3(s + 1)}.$$

Multiplying through by $s^3(s + 1)$ and equating like terms yields the system of equations

$$\begin{cases} A + D = 1 \\ A + B = 0 \\ B + C = 0 \\ C = 2. \end{cases}$$

Solving simultaneously gives us $A = 2$, $B = -2$, $C = 2$, and $D = -1$. Therefore we seek a function $y(t)$ whose Laplace transform is

$$\frac{2}{s} - \frac{2}{s^2} + \frac{2}{s^3} - \frac{1}{s + 1}.$$

We have $\mathcal{L}[e^{-t}] = 1/(s + 1)$ so that

$$\mathcal{L}[-e^{-t}] = -\mathcal{L}[e^{-t}] = -\frac{1}{s + 1}.$$

Also, using the formula from Exercise 5, we have

$$\mathcal{L}[t^2] = \frac{2}{s^3}, \quad \mathcal{L}[t] = \frac{1}{s^2}, \quad \text{and} \quad \mathcal{L}[1] = \frac{1}{s}$$

so that

$$\mathcal{L}[t^2 - 2t + 2] = \mathcal{L}[t^2] - 2\mathcal{L}[t] + 2\mathcal{L}[1] = \frac{2}{s^3} - \frac{2}{s^2} + \frac{2}{s}.$$

Therefore, $y(t) = t^2 - 2t + 2 - e^{-t}$ is the desired function.

(d) Since $y(0) = 2 - e^0 = 1$, $y(t)$ satisfies the given initial condition. Also,

$$\frac{dy}{dt} = 2t - 2 + e^{-t}$$

and

$$-y + t^2 = -t^2 + 2t - 2 + e^{-t} + t^2 = 2t - 2 + e^{-t}$$

so our solution also satisfies the differential equation.

25. First take Laplace transforms of both sides of the equation

$$\mathcal{L}\left[\frac{dy}{dt}\right] = 2\mathcal{L}[y] + 2\mathcal{L}[e^{-3t}]$$

and use the rules to simplify, obtaining

$$s\mathcal{L}[y] - y(0) = 2\mathcal{L}[y] + \frac{2}{s+3}$$

$$(s - 2)\mathcal{L}[y] = y(0) + \frac{2}{s+3}$$

$$\mathcal{L}[y] = \frac{y(0)}{s-2} + \frac{2}{(s-2)(s+3)}.$$

Next note that

$$\mathcal{L}[y(0)e^{2t}] = y(0)/(s-2).$$

For the other summand, first simplify using partial fractions,

$$\frac{2}{(s-2)(s+3)} = \frac{A}{s-2} + \frac{B}{s+3}.$$

Putting the right-hand side over a common denominator gives $A(s+3) + B(s-2) = 2$, which can be written as $(A + B)s + (3A - 2B) = 2$. This yields $A + B = 0$ and $3A - 2B = 2$. Hence $B = -2/5$ and $A = 2/5$, and

$$\frac{2}{(s-2)(s+3)} = \frac{2/5}{s-2} - \frac{2/5}{s+3}.$$

Now, $\mathcal{L}[e^{2t}] = 1/(s-2)$ and $\mathcal{L}[e^{-3t}] = 1/(s+3)$ so

$$\mathcal{L}[y] = \frac{y(0)}{s-2} + \frac{2}{5}\frac{1}{s-2} - \frac{2}{5}\frac{1}{s+3}.$$

Hence,

$$y(t) = y(0)e^{2t} + \frac{2}{5}e^{2t} - \frac{2}{5}e^{-3t}.$$

The first two terms can be combined into one, giving

$$y(t) = ce^{2t} - \frac{2}{5}e^{-3t},$$

where $c = y(0) + 2/5$.

27. As always the first step must be to take Laplace transform of both sides of the differential equation, giving

$$\mathcal{L}\left[\frac{dy}{dt}\right] = \mathcal{L}[y^2].$$

Simplifying, we obtain

$$s\mathcal{L}[y] - 1 = \mathcal{L}[y^2].$$

To solve for $\mathcal{L}[y]$ we must come up with an expression for $\mathcal{L}[y^2]$ in terms of $\mathcal{L}[y]$. This is not so easy! In particular, there is no easy way to simplify

$$\mathcal{L}[y^2] = \int_0^\infty y^2 e^{-st}\, dt$$

since we do not have a rule for the Laplace transform of a product.

EXERCISES FOR SECTION 6.2

1.   **(a)** The function $g_a(t) = 1$ precisely when $u_a(t) = 0$, and $g_a(t) = 0$ precisely when $u_a(t) = 1$, so

$$g_a(t) = 1 - u_a(t).$$

(b) We can compute the Laplace transform of $g_a(t)$ from the definition

$$\mathcal{L}[g_a] = \int_0^a 1e^{-st}\, dt = -\frac{e^{-as}}{s} + \frac{e^{-0s}}{s} = \frac{1}{s} - \frac{e^{-as}}{s}.$$

Alternately, we can use the table

$$\mathcal{L}[g_a] = \mathcal{L}[1 - u_a(t)] = \frac{1}{s} - \frac{e^{-as}}{s}.$$

3.

$$\mathcal{L}[g_a(t)] = \int_0^\infty g_a(t)\, e^{-st}\, dt$$

$$= \int_0^a \frac{t}{a} e^{-st}\, dt + \int_a^\infty e^{-st}\, dt$$

Using integration by parts with $u = t$ and $dv = e^{-st}dt$, we have $du = dt$, $v = -e^{-st}/s$ and

$$\int_0^a \frac{t}{a} e^{-st}\, dt = \frac{1}{a} \int_0^a t e^{-st}\, dt$$

$$= \frac{1}{a} \left(-\frac{t e^{-st}}{s} \Big|_0^a - \int_0^a -\frac{e^{-st}}{s}\, dt \right)$$

$$= \frac{1}{a} \left(-\frac{a e^{-as}}{s} - \frac{1}{s^2} e^{-st} \Big|_0^a \right)$$

$$= \frac{1}{a} \left(-\frac{a e^{-as}}{s} - \frac{1}{s^2} (e^{-as} - 1) \right)$$

$$= -\frac{e^{-as}}{s} - \frac{1}{as^2} (e^{-as} - 1).$$

Also,

$$\int_a^\infty e^{-st}\, dt = \lim_{b \to \infty} \int_a^b e^{-st}\, dt$$

$$= \lim_{b \to \infty} -\frac{1}{s} e^{-st} \Big|_a^b$$

$$= \lim_{b \to \infty} -\frac{1}{s} \left(e^{-sb} - e^{-as} \right)$$

$$= \frac{1}{s} e^{-as}.$$

Therefore,

$$\mathcal{L}[g_a(t)] = -\frac{e^{-as}}{s} - \frac{1}{as^2}\left(e^{-as} - 1\right) + \frac{1}{s}e^{-as}$$

$$= \frac{1}{as^2}\left(1 - e^{-as}\right).$$

5. First use partial fractions to write

$$\frac{1}{(s-1)(s-2)} = \frac{A}{s-1} + \frac{B}{s-2}.$$

Putting the right-hand side over a common denominator yields $As - 2A + Bs - B = 1$ which can be written as $(A + B)s + (-2A - B) = 1$. Thus, $A + B = 0$, and $-2A - B = 1$. Solving for A and B yields $A = -1$ and $B = 1$, so

$$\frac{1}{(s-1)(s-2)} = \frac{-1}{s-1} + \frac{1}{s-2}.$$

Now, as above

$$\mathcal{L}[u_3(t)e^{2(t-3)}] = \frac{e^{-3s}}{s-2}$$

and

$$\mathcal{L}[u_3(t)e^{t-3}] = \frac{e^{-3s}}{s-1}$$

and the desired function is

$$u_3(t)\left(e^{2(t-3)} - e^{(t-3)}\right).$$

7. Using partial fractions, we get

$$\frac{14}{(3s+2)(s-4)} = \frac{A}{3s+2} + \frac{B}{s-4}.$$

Hence, we must have $As - 4A + 3Bs + 2B = 14$, which can be written as

$$(A + 3B)s + (-4A + 2B) = 14.$$

Therefore, $A + 3B = 0$, and $-4A + 2B = 14$. Solving for A and B yields $A = -3$ and $B = 1$, so

$$\frac{14}{(3s+2)(s-4)} = \frac{1}{s-4} - \frac{3}{3s+2} = \frac{1}{s-4} - \frac{1}{s+2/3}.$$

Applying the rules

$$\mathcal{L}[u_1(t)e^{4(t-1)}] = \frac{e^{-s}}{s-4}$$

and

$$\mathcal{L}[u_1(t)e^{-\frac{2}{3}(t-1)}] = \frac{e^{-s}}{s+2/3},$$

the desired function is

$$y(t) = u_1(t)\left(e^{4(t-1)} - e^{-\frac{2}{3}(t-1)}\right).$$

9. Taking the Laplace transform of both sides of the equation, we have

$$\mathcal{L}\left[\frac{dy}{dt}\right] + 9\mathcal{L}[y] = \mathcal{L}[u_5(t)],$$

which is equivalent to

$$s\mathcal{L}[y] - y(0) + 9\mathcal{L}[y] = \frac{e^{-5s}}{s}.$$

Since $y(0) = -2$, we have

$$s\mathcal{L}[y] + 2 + 9\mathcal{L}[y] = \frac{e^{-5s}}{s},$$

which yields

$$\mathcal{L}[y] = \frac{-2}{s+9} + \frac{e^{-5s}}{s(s+9)}.$$

Using the partial fractions decomposition

$$\frac{1}{s(s+9)} = \frac{1/9}{s} - \frac{1/9}{s+9},$$

we see that

$$\mathcal{L}[y] = \frac{-2}{s+9} + \frac{1}{9}\left(\frac{e^{-5s}}{s}\right) - \frac{1}{9}\left(\frac{e^{-5s}}{s+9}\right).$$

Taking the inverse of the Laplace transform, we obtain

$$y(t) = -2e^{-9t} + \frac{1}{9}u_5(t) - \frac{1}{9}u_5(t)e^{-9(t-5)}$$

$$= -2e^{-9t} + \frac{1}{9}u_5(t)\left(1 - e^{-9(t-5)}\right).$$

To check our answer, we compute

$$\frac{dy}{dt} = 18e^{-9t} + \frac{1}{9}\frac{du_5}{dt}\left(1 - e^{-9(t-5)}\right) + \frac{1}{9}u_5(t)\left(9e^{-9(t-5)}\right),$$

and since $du_5/dt = 0$ except at $t = 5$ (where it is undefined),

$$\frac{dy}{dt} + 9y = 18e^{-9t} + u_5(t)e^{-9(t-5)} + 9\left(-2e^{-9t} + \frac{1}{9}u_5(t)\left(1 - e^{-9(t-5)}\right)\right)$$

$$= u_5(t).$$

Hence, our $y(t)$ satisfies the differential equation except when $t = 5$. (We cannot expect $y(t)$ to satisfy the differential equation at $t = 5$ because the differential equation is not continuous there.) Note that $y(t)$ also satisfies the initial condition $y(0) = -2$.

11. Taking the Laplace transform of both sides of the equation, we obtain

$$\mathcal{L}\left[\frac{dy}{dt}\right] = \mathcal{L}[-y] + \mathcal{L}[u_2(t)e^{-2(t-2)}],$$

which is equivalent to

$$s\mathcal{L}[y] - y(0) = -\mathcal{L}[y] + \frac{e^{-2s}}{s+2}$$

(using linearity of the Laplace transform and the formula
$$\text{If } \mathcal{L}[f] = F(s) \text{ then } \mathcal{L}[u_a(t)f(t-a)] = e^{-as}F(s)$$
where $f(t) = e^{-2t}$ and $a = 2$.)
Substituting the initial condition yields

$$s\mathcal{L}[y] - 1 = -\mathcal{L}[y] + \frac{e^{-2s}}{s+2}$$

so that

$$\mathcal{L}[y] = \frac{1}{s+1} + \frac{e^{-2s}}{(s+1)(s+2)}.$$

By partial fractions, we know that

$$\frac{1}{s+1} - \frac{1}{s+2} = \frac{1}{(s+1)(s+2)},$$

so we have

$$\frac{e^{-2s}}{(s+1)(s+2)} = e^{-2s}\left(\frac{1}{s+1} - \frac{1}{s+2}\right) = \frac{e^{-2s}}{s+1} - \frac{e^{-2s}}{s+2}.$$

Taking the inverse of the Laplace transform yields

$$y(t) = e^{-t} + u_2(t)e^{-(t-2)} - u_2(t)e^{-2(t-2)}$$

$$= e^{-t} + u_2(t)\left(e^{-(t-2)} - e^{-2(t-2)}\right).$$

To check our answer, we compute

$$\frac{dy}{dt} = -e^{-t} + \frac{du_2}{dt}\left(e^{-(t-2)} - e^{-2(t-2)}\right) + u_2(t)\left(-e^{-(t-2)} + 2e^{-2(t-2)}\right),$$

and since $du_2/dt = 0$ except at $t = 2$ (where it is undefined),

$$\frac{dy}{dt} + y = -e^{-t} + u_2(t)\left(-e^{-(t-2)} + 2e^{-2(t-2)}\right) + e^{-t} + u_2(t)\left(e^{-(t-2)} - e^{-2(t-2)}\right)$$

$$= u_2(t)e^{-2(t-2)}.$$

Hence, our $y(t)$ satisfies the differential equation except when $t = 2$. (We cannot expect $y(t)$ to satisfy the differential equation at $t = 2$ because the differential equation is not continuous there.) Note that $y(t)$ also satisfies the initial condition $y(0) = 1$.

13. Taking the Laplace transform of both sides of the equation, we obtain

$$\mathcal{L}\left[\frac{dy}{dt}\right] = -\mathcal{L}[y] + \mathcal{L}[u_1(t)(t-1)],$$

which is equivalent to

$$s\mathcal{L}[y] - y(0) = -\mathcal{L}[y] + \frac{e^{-s}}{s^2}.$$

Substituting the initial condition yields

$$s\mathcal{L}[y] - 2 = -\mathcal{L}[y] + \frac{e^{-s}}{s^2}$$

so that

$$\mathcal{L}[y] = \frac{e^{-s}}{s^2(s+1)} + \frac{2}{s+1}.$$

Using the technique of partial fractions, we write

$$\frac{1}{s^2(s+1)} = \frac{A}{s} + \frac{B}{s^2} + \frac{C}{s+1}.$$

Putting the right-hand side over a common denominator gives us $As(s+1) + B(s+1) + Cs^2 = 1$ which can be written as $(A+C)s^2 + (A+B)s + B = 1$. So $A+C = 0$, $A+B = 0$, and $B = 1$. Thus $A = -1$ and $C = 1$, and

$$\frac{1}{s^2(s+1)} = \frac{-1}{s} + \frac{1}{s^2} + \frac{1}{s+1}.$$

Taking the inverse of the Laplace transform gives us

$$y(t) = \mathcal{L}^{-1}\left[\frac{e^{-s}}{s^2(s+1)}\right] + \mathcal{L}^{-1}\left[\frac{2}{s+1}\right]$$

$$= -\mathcal{L}^{-1}\left[\frac{e^{-s}}{s}\right] + \mathcal{L}^{-1}\left[\frac{e^{-s}}{s^2}\right] + \mathcal{L}^{-1}\left[\frac{e^{-s}}{s+1}\right] + \mathcal{L}^{-1}\left[\frac{2}{s+1}\right]$$

$$= -u_1(t) + u_1(t)(t-1) + u_1(t)e^{-(t-1)} + 2e^{-t}$$

$$= u_1(t)\left((t-2) + e^{-(t-1)}\right) + 2e^{-t}.$$

To check our answer, we compute

$$\frac{dy}{dt} = \frac{du_1}{dt}\left((t-2) + e^{-(t-1)}\right) + u_1(t)\left(1 - e^{-(t-1)}\right) - 2e^{-t},$$

and since $du_1/dt = 0$ except at $t = 1$ (where it is undefined),

$$\frac{dy}{dt} + y = u_1(t)\left(1 - e^{-(t-1)}\right) - 2e^{-t} + u_1(t)\left((t-2) + e^{-(t-1)}\right) + 2e^{-t}$$

$$= u_1(t) + u_1(t)(t-2)$$

$$= u_1(t)(t-1).$$

Hence, our $y(t)$ satisfies the differential equation except when $t = 1$. (We cannot expect $y(t)$ to satisfy the differential equation at $t = 1$ because the differential equation is not continuous there.) Note that $y(t)$ also satisfies the initial condition $y(0) = 2$.

15. Taking the Laplace transform of both sides of the equation, we have

$$\mathcal{L}\left[\frac{dy}{dt}\right] = -\mathcal{L}[y] + \mathcal{L}[u_a(t)],$$

which is equivalent to

$$s\mathcal{L}[y] - y(0) = -\mathcal{L}[y] + \frac{e^{-as}}{s}.$$

Solving for $\mathcal{L}[y]$ yields

$$\mathcal{L}[y] = \frac{e^{-as}}{s(s+1)} + \frac{y(0)}{s+1}.$$

Using the partial fractions decomposition

$$\frac{1}{s(s+1)} = \frac{1}{s} - \frac{1}{s+1},$$

we get

$$\mathcal{L}[y] = \frac{e^{-as}}{s} - \frac{e^{-as}}{s+1} + \frac{y(0)}{s+1}.$$

Taking the inverse Laplace transform, we obtain

$$y(t) = u_a(t) - u_a(t)e^{-(t-a)} + y(0)e^{-t}$$

$$= u_a(t)\left(1 - e^{-(t-a)}\right) + y(0)e^{-t}.$$

To check our answer, we compute

$$\frac{dy}{dt} = \frac{du_a}{dt}\left(1 - e^{-(t-a)}\right) + u_a(t)e^{-(t-a)} - y(0)e^{-t}$$

and since $du_a/dt = 0$ except at $t = a$ (where it is undefined),

$$\frac{dy}{dt} + y = u_a(t)e^{-(t-a)} - y(0)e^{-t} + u_a(t)\left(1 - e^{-(t-a)}\right) + y(0)e^{-t}$$

$$= u_a(t).$$

Hence, our $y(t)$ satisfies the differential equation except when $t = a$. (We cannot expect $y(t)$ to satisfy the differential equation at $t = a$ because the differential equation is not continuous there.)

17. From the formula in Exercise 16 we need only compute

$$\int_0^2 w(t)e^{-st}\,dt.$$

To compute this integral, we write

$$\int_0^2 w(t)e^{-st}\,dt = \int_0^1 e^{-st}\,dt - \int_1^2 e^{-st}\,dt$$

$$= \frac{e^{-st}}{-s}\bigg|_0^1 - \frac{e^{-st}}{-s}\bigg|_1^2$$

$$= \left(-\frac{e^{-s}}{s} + \frac{1}{s}\right) - \left(-\frac{e^{-2s}}{s} + \frac{e^{-s}}{s}\right)$$

$$= \frac{1}{s} - 2\frac{e^{-s}}{s} + \frac{e^{-2s}}{s}.$$

Using the formula of Exercise 16 gives

$$\mathcal{L}[w] = \frac{1}{1 - e^{-2s}}\left(\frac{1}{s} - 2\frac{e^{-s}}{s} + \frac{e^{-2s}}{s}\right) = \frac{1 - 2e^{-s} + e^{-2s}}{s(1 - e^{-2s})}.$$

This can be simplified to

$$\mathcal{L}[w] = \frac{(1 - e^{-s})^2}{s(1 - e^{-s})(1 + e^{-s})} = \frac{1 - e^{-s}}{s(1 + e^{-s})}.$$

19. **(a)** Taking the Laplace transform of both sides of the equations yields

$$\mathcal{L}\left[\frac{dy}{dt}\right] = -\mathcal{L}[y] + \mathcal{L}[w(t)],$$

which is equivalent to

$$s\mathcal{L}[y] - y(0) = -\mathcal{L}[y] + \frac{1 - e^{-s}}{s(1 + e^{-s})},$$

using linearity of the Laplace transform and the results of Exercise 17. Since $y(0) = 0$, solving for $\mathcal{L}[y]$ gives

$$\mathcal{L}[y] = \frac{1 - e^{-s}}{s(s + 1)(1 + e^{-s})}.$$

(b) The function $w(t)$ is alternatively 1 and -1. While $w(t) = 1$, the solution decays exponentially toward $y = 1$. When $w(t)$ changes to -1, the solution then decays toward $y = -1$.
In fact, there is a periodic solution with initial condition $y(0) = (1 - e)/(1 + e) \approx -0.462$, and our solution tends toward this periodic solution as $t \to \infty$.

EXERCISES FOR SECTION 6.3

1. We use integration by parts twice to compute

$$\mathcal{L}[\sin \omega t] = \int_0^\infty \sin \omega t \, e^{-st} \, dt.$$

First, letting $u = \sin \omega t$ and $dv = e^{-st}\, dt$, we get

$$\mathcal{L}[\sin \omega t] = \sin \omega t \left. \frac{e^{-st}}{-s} \right|_0^\infty - \int_0^\infty \frac{e^{-st}}{-s} \omega \cos \omega t\, dt$$

$$= \lim_{b \to \infty} \left[\frac{e^{-st}}{-s} \sin \omega t \Big|_0^b \right] + \frac{\omega}{s} \int_0^\infty e^{-st} \cos \omega t\, dt$$

$$= \frac{\omega}{s} \int_0^\infty e^{-st} \cos \omega t\, dt,$$

since the limit of $e^{-sb} \sin \omega b$ is 0 as $b \to \infty$ and $s > 0$.

Using integration by parts on

$$\int_0^\infty e^{-st} \cos \omega t\, dt,$$

with $u = \cos \omega t$ and $dv = e^{-st}\, dt$, we get

$$\int_0^\infty e^{-st} \cos \omega t\, dt = \left. \frac{e^{-st}}{-s} \cos \omega t \right|_0^\infty - \int_0^\infty \frac{e^{-st}}{-s} (-\omega \sin \omega t)\, dt$$

$$= \lim_{b \to \infty} \left[\frac{e^{-st}}{-s} \cos \omega t \Big|_0^b \right] - \frac{\omega}{s} \int_0^\infty e^{-st} \sin \omega t\, dt$$

$$= \lim_{b \to \infty} \left[\frac{e^{-sb}}{-s} \cos \omega b \right] + \frac{1}{s} - \frac{\omega}{s} \int_0^\infty e^{-st} \sin \omega t\, dt$$

$$= \frac{1}{s} - \frac{\omega}{s} \int_0^\infty e^{-st} \sin \omega t\, dt,$$

since the limit of $e^{-sb} \cos \omega b$ is 0 as $b \to \infty$ and $s > 0$.

Thus,

$$\int_0^\infty \sin \omega t\, e^{-st}\, dt = \frac{\omega}{s} \int_0^\infty e^{-st} \cos \omega t\, dt$$

$$= \frac{\omega}{s} \left(\frac{1}{s} - \frac{\omega}{s} \int_0^\infty \sin \omega t\, e^{-st}\, dt \right)$$

$$= \frac{\omega}{s^2} - \frac{\omega^2}{s^2} \int_0^\infty \sin \omega t\, e^{-st}\, dt.$$

So

$$\frac{s^2 + \omega^2}{s^2} \int_0^\infty \sin \omega t\, e^{-st}\, dt = \frac{\omega}{s^2},$$

and

$$\int_0^\infty \sin \omega t\, e^{-st}\, dt = \frac{\omega}{s^2 + \omega^2}.$$

3. We need to compute

$$\mathcal{L}[e^{at}\sin\omega t] = \int_0^\infty e^{at}\sin\omega t\, e^{-st}\, dt = \int_0^\infty \sin\omega t\, e^{-(s-a)t}\, dt.$$

We can do this using integration by parts twice and ending up with $\mathcal{L}[e^{at}\sin\omega t]$ on both sides of the equation. Alternately, if we let $r = s - a$, then

$$\int_0^\infty \sin\omega t\, e^{-(s-a)t}\, dt = \int_0^\infty \sin\omega t\, e^{-rt}\, dt$$

The integral on the right is the Laplace transform of $\sin\omega t$ with r as the new independent variable. From Exercise 1, we know

$$\int_0^\infty \sin\omega t\, e^{-rt}\, dt = \frac{\omega}{r^2 + \omega^2}.$$

Substituting back we have

$$\mathcal{L}[e^{at}\sin\omega t] = \frac{\omega}{(s-a)^2 + \omega^2}.$$

5. Using the formula

$$\mathcal{L}\left[\frac{d^2 y}{dt^2}\right] = s^2\mathcal{L}[y] - y'(0) - sy(0),$$

and the linearity of the Laplace transform, we get that

$$s^2\mathcal{L}[y] - y'(0) - sy(0) + \omega^2\mathcal{L}[y] = 0.$$

Substituting the initial conditions and solving for $\mathcal{L}[y]$ gives

$$\mathcal{L}[y] = \frac{s}{s^2 + \omega^2}.$$

7. Since

$$\mathcal{L}[\sin\omega t] = \frac{\omega}{s^2 + \omega^2},$$

we can compute that

$$\frac{d}{d\omega}\mathcal{L}[\sin\omega t] = \frac{s^2 - \omega^2}{(s^2 + \omega^2)^2},$$

but

$$\frac{d}{d\omega}\mathcal{L}[\sin\omega t] = \mathcal{L}\left[\frac{d}{d\omega}\sin\omega t\right] = \mathcal{L}[t\cos\omega t].$$

So

$$\mathcal{L}[t\cos\omega t] = \frac{s^2 - \omega^2}{(s^2 + \omega^2)^2}.$$

9. From Exercise 8, we know that

$$\mathcal{L}[te^{at}] = \frac{1}{(s-a)^2}.$$

Differentiating both sides of this formula with respect to a gives

$$\frac{d}{da}\mathcal{L}[te^{at}] = \mathcal{L}\left[\frac{d}{da}te^{ta}\right] = \mathcal{L}[t^2 e^{at}]$$

while

$$\frac{d}{da}\frac{1}{(s-a)^2} = \frac{2}{(s-a)^3}.$$

Hence,

$$\mathcal{L}[t^2 e^{at}] = \frac{2}{(s-a)^3}.$$

11. In this case, $b = 2$, and $(s + b/2)^2 = (s+1)^2 = s^2 + 2s + 1$, so $s^2 + 2s + 10 = (s+1)^2 + 3^2$.

13. In this case, $b = 1$, and $(s + b/2)^2 = (s+1/2)^2 = s^2 + s + 1/4$, so $s^2 + s + 1 = (s+1/2)^2 + 3/4 = (s+1/2)^2 + (\sqrt{3}/2)^2$.

15. In Exercise 11, we completed the square and obtained $s^2 + 2s + 10 = (s+1)^2 + 3^2$, so

$$\mathcal{L}^{-1}\left[\frac{1}{s^2 + 2s + 10}\right] = \mathcal{L}^{-1}\left[\frac{1}{(s+1)^2 + 3^2}\right]$$

$$= \frac{1}{3}\mathcal{L}^{-1}\left[\frac{3}{(s+1)^2 + 3^2}\right]$$

$$= \frac{1}{3}e^{-t}\sin 3t.$$

17. In Exercise 13, we completed the square and obtained $s^2 + s + 1 = (s+1/2)^2 + (\sqrt{3}/2)^2$, so

$$\frac{2s+3}{s^2+s+1} = \frac{2s+3}{(s+1/2)^2 + (\sqrt{3}/2)^2}.$$

We want to put this fraction in the right form so that we can use the formulas for $\mathcal{L}[e^{at}\cos\omega t]$ and $\mathcal{L}[e^{at}\sin\omega t]$. We see that

$$\frac{2s+3}{(s+1/2)^2 + (\sqrt{3}/2)^2} = \frac{2s+1}{(s+1/2)^2 + (\sqrt{3}/2)^2} + \frac{2}{(s+1/2)^2 + (\sqrt{3}/2)^2}$$

$$= \frac{2(s+1/2)}{(s+1/2)^2 + (\sqrt{3}/2)^2} + \frac{(4/\sqrt{3})(\sqrt{3}/2)}{(s+1/2)^2 + (\sqrt{3}/2)^2}.$$

So

$$\mathcal{L}^{-1}\left[\frac{2s+3}{s^2+s+1}\right] = 2\mathcal{L}^{-1}\left[\frac{(s+1/2)}{(s+1/2)^2 + (\sqrt{3}/2)^2}\right] + \frac{4}{\sqrt{3}}\mathcal{L}^{-1}\left[\frac{\sqrt{3}/2}{(s+1/2)^2 + (\sqrt{3}/2)^2}\right]$$

$$= 2e^{-t/2}\cos\left(\frac{\sqrt{3}}{2}t\right) + \frac{4}{\sqrt{3}}e^{-t/2}\sin\left(\frac{\sqrt{3}}{2}t\right).$$

19. We compute

$$\mathcal{L}\left[e^{(a+ib)t}\right] = \int_0^\infty e^{(a+ib)t} e^{-st} \, dt$$

$$= \int_0^\infty e^{-(s-(a+ib))t} \, dt$$

$$= -\frac{1}{s-(a+ib)} \left(\lim_{u\to\infty}\left[e^{-(s-a)u}e^{-ibu}\right] - 1\right).$$

The limit is zero as long as $s > a$. Hence,

$$\mathcal{L}\left[e^{(a+ib)t}\right] = \frac{1}{s-(a+ib)}$$

if $s > a$ and undefined otherwise. This is the same formula as for real exponentials. It can also be written

$$\mathcal{L}\left[e^{(a+ib)t}\right] = \frac{s-a+ib}{(s-a)^2 + b^2}.$$

21. We recall that

$$e^{at}\cos\omega t = \text{Re}(e^{(a+ib)t}).$$

So

$$\mathcal{L}[e^{at}\cos\omega t] = \text{Re}(\mathcal{L}[e^{(a+ib)t}])$$

$$= \text{Re}\left(\frac{s-a+i\omega}{(s-a)^2+\omega^2}\right)$$

$$= \frac{s-a}{(s-a)^2+\omega^2}.$$

Similarly,

$$\mathcal{L}[e^{at}\sin\omega t] = \text{Im}(\mathcal{L}[e^{(a+ib)t}])$$

$$= \text{Im}\left(\frac{s-a+i\omega}{(s-a)^2+\omega^2}\right)$$

$$= \frac{\omega}{(s-a)^2+\omega^2}.$$

23. Using the quadratic formula, we see that the roots of $s^2 + 2s + 10 = 0$ are $s = -1 \pm 3i$. Thus $s^2 + 2s + 10 = (s+1+3i)(s+1-3i)$. So we want to find A and B so that

$$\frac{1}{s^2+2s+10} = \frac{A}{s+1+3i} + \frac{B}{s+1-3i}.$$

So, finding common denominators (that is, usual partial fractions only with complex numbers) gives

$$\begin{cases} A + B = 0 \\ A + B + 3i(-A + B) = 1. \end{cases}$$

Solving, we get $A = i/6$ and $B = -i/6$, so

$$\frac{1}{s^2 + 2s + 10} = \frac{i/6}{s + 1 + 3i} + \frac{-i/6}{s + 1 - 3i}.$$

Thus

$$\mathcal{L}^{-1}\left[\frac{1}{s^2 + 2s + 10}\right] = \mathcal{L}^{-1}\left[\frac{i/6}{s + 1 + 3i} + \frac{-i/6}{s + 1 - 3i}\right]$$

$$= \frac{i}{6}e^{(-1-3i)t} - \frac{i}{6}e^{(-1+3i)t}$$

$$= \frac{i}{6}\left(e^{-t}\cos(-3t) + ie^{-t}\sin(-3t)\right) - \frac{i}{6}\left(e^{-t}\cos 3t + ie^{-t}\sin 3t\right)$$

$$= -\frac{i}{6}\left(2ie^{-t}\sin 3t\right)$$

$$= \frac{1}{3}e^{-t}\sin 3t.$$

25. Using the quadratic formula, the roots of the denominator are $(-1 \pm i\sqrt{3})/2$. Hence, we can factor the denominator into

$$\left(s - \left(\tfrac{-1+i\sqrt{3}}{2}\right)\right)\left(s - \left(\tfrac{-1-i\sqrt{3}}{2}\right)\right).$$

We then do the partial fractions decomposition

$$\frac{2s + 3}{s^2 + s + 1} = \frac{A}{s - \left(\tfrac{-1+i\sqrt{3}}{2}\right)} + \frac{B}{s - \left(\tfrac{-1-i\sqrt{3}}{2}\right)},$$

which gives rise to the equations

$$\begin{cases} A + B = 2 \\ \left(\tfrac{1+i\sqrt{3}}{2}\right)A + \left(\tfrac{1-i\sqrt{3}}{2}\right)B = 3. \end{cases}$$

Solving yields $A = 1 - \frac{2}{\sqrt{3}}i$ and $B = 1 + \frac{2}{\sqrt{3}}i$. So

$$\frac{2s + 3}{s^2 + s + 1} = \frac{1 - \frac{2}{\sqrt{3}}i}{s - \left(\tfrac{-1+i\sqrt{3}}{2}\right)} + \frac{1 + \frac{2}{\sqrt{3}}i}{s - \left(\tfrac{-1-i\sqrt{3}}{2}\right)}.$$

Taking inverse Laplace transforms of the right-hand side gives

$$\left(1 - \tfrac{2}{\sqrt{3}}i\right) e^{(-1+i\sqrt{3})t/2} + \left(1 + \tfrac{2}{\sqrt{3}}i\right) e^{(-1-i\sqrt{3})t/2}.$$

Using Euler's formula to replace the complex exponentials and simplifying yields

$$2e^{-t/2} \cos\left(\tfrac{\sqrt{3}}{2}t\right) + \tfrac{4}{\sqrt{3}}e^{-t/2} \sin\left(\tfrac{\sqrt{3}}{2}t\right).$$

27. **(a)** Taking the Laplace transform of both sides of the equation, we obtain

$$\mathcal{L}\left[\frac{d^2y}{dt^2}\right] + 4\mathcal{L}[y] = \frac{8}{s},$$

and using the fact that $\mathcal{L}[d^2y/dt^2] = s^2\mathcal{L}[y] - sy(0) - y'(0)$, we have

$$(s^2 + 4)\mathcal{L}[y] - sy(0) - y'(0) = \frac{8}{s}.$$

(b) Substituting the initial conditions yields

$$(s^2 + 4)\mathcal{L}[y] - 11s - 5 = \frac{8}{s},$$

and solving for $\mathcal{L}[y]$ we get

$$\mathcal{L}[y] = \frac{11s + 5}{s^2 + 4} + \frac{8}{s(s^2 + 4)}.$$

The partial fractions decomposition of $8/(s(s^2 + 4))$ is

$$\frac{8}{s(s^2 + 4)} = \frac{A}{s} + \frac{Bs + C}{s^2 + 4}.$$

Putting the right-hand side over a common denominator gives us

$$(A + B)s^2 + Cs + 4A = 8,$$

and consequently, $A = 2$, $B = -2$, and $C = 0$. In other words,

$$\frac{8}{s(s^2 + 4)} = \frac{2}{s} + \frac{-2s}{s^2 + 4}.$$

We obtain

$$\mathcal{L}[y] = \frac{2}{s} + \frac{9s + 5}{s^2 + 4}.$$

(c) To take the inverse Laplace transform, we rewrite $\mathcal{L}[y]$ in the form

$$\mathcal{L}[y] = \frac{2}{s} + 9\left(\frac{s}{s^2 + 4}\right) + \frac{5}{2}\left(\frac{2}{s^2 + 4}\right).$$

Therefore, $y(t) = 2 + 9\cos 2t + \tfrac{5}{2}\sin 2t$.

29. **(a)** Taking the Laplace transform of both sides of the equation, we obtain

$$\mathcal{L}\left[\frac{d^2 y}{dt^2}\right] - 4\mathcal{L}\left[\frac{dy}{dt}\right] + 5\mathcal{L}[y] = \frac{2}{s-1},$$

and using the formulas for $\mathcal{L}[dy/dt]$ and $\mathcal{L}[d^2 y/dt^2]$ in terms of $\mathcal{L}[y]$, we have

$$(s^2 - 4s + 5)\mathcal{L}[y] - sy(0) - y'(0) + 4y(0) = \frac{2}{s-1}.$$

(b) Substituting the initial conditions yields

$$(s^2 - 4s + 5)\mathcal{L}[y] - 3s + 11 = \frac{1}{s-2},$$

and solving for $\mathcal{L}[y]$ we get

$$\mathcal{L}[y] = \frac{3s - 11}{s^2 - 4s + 5} + \frac{1}{(s-1)(s^2 - 4s + 5)}.$$

Using the partial fractions decomposition

$$\frac{1}{(s-1)(s^2 - 4s + 5)} = \frac{1}{s-1} + \frac{-s+3}{s^2 - 4s + 5},$$

we obtain

$$\mathcal{L}[y] = \frac{1}{s-1} + \frac{2s - 8}{s^2 - 4s + 5}.$$

(c) In order to compute the inverse Laplace transform, we first write

$$s^2 - 4s + 5 = (s-2)^2 + 1$$

by completing the square, and then we write

$$\frac{2s - 8}{s^2 - 4s + 5} = \frac{2(s-2)}{(s-2)^2 + 1} - \frac{4}{(s-2)^2 + 1}.$$

Taking the inverse Laplace transform, we have

$$y(t) = e^t + 2e^{2t}\cos t - 4e^{2t}\sin t.$$

31. **(a)** Note that this is resonant forcing of an undamped oscillator. We take the Laplace transform of both sides

$$\mathcal{L}\left[\frac{d^2 y}{dt^2}\right] + 4\mathcal{L}[y] = \mathcal{L}[\cos 2t]$$

and obtain

$$s^2 \mathcal{L}[y] + 2s + 4\mathcal{L}[y] = \frac{s}{s^2 + 4}.$$

(b) Solving for $\mathcal{L}[y]$, we get

$$\mathcal{L}[y] = -\frac{2s}{s^2 + 4} + \frac{s}{(s^2 + 4)^2}.$$

(c) To take the inverse Laplace transform, we note that

$$\mathcal{L}^{-1}\left[-\frac{2s}{s^2+4}\right] = -2\cos 2t$$

and

$$\mathcal{L}^{-1}\left[\frac{s}{(s^2+4)^2}\right] = \frac{1}{4}\mathcal{L}^{-1}\left[\frac{4s}{(s^2+4)^2}\right] = \frac{t}{4}\sin 2t.$$

So

$$y(t) = -2\cos 2t + \frac{t}{4}\sin 2t,$$

which is of the form we would expect for a resonant response.

33. **(a)** First take the Laplace transform of both sides of the equation

$$\mathcal{L}\left[\frac{d^2y}{dt^2}\right] + 3\mathcal{L}[y] = \mathcal{L}[w(t)].$$

We need to compute $\mathcal{L}[w(t)]$. One way to do this is to use the definition

$$\mathcal{L}[w(t)] = \int_0^\infty w(t)e^{-st}\,dt = \int_0^1 te^{-st}\,dt + \int_1^\infty e^{-st}\,dt$$

Evaluating the first integral by parts and the second integral directly yields

$$\mathcal{L}[w(t)] = -\frac{e^{-s}}{s} - \frac{e^{-s}}{s^2} + \frac{1}{s^2} + \frac{e^{-s}}{s} = \frac{1-e^{-s}}{s^2}.$$

(We could also write $w(t) = t - (t-1)u_1(t)$ and use the table.)
Hence, after we transform the equation, we get

$$s^2\mathcal{L}[y] - 2s + 3\mathcal{L}[y] = \frac{1-e^{-s}}{s^2}.$$

(b) Solving for $\mathcal{L}[y]$, we obtain

$$\mathcal{L}[y] = \frac{2s}{s^2+3} + \frac{1-e^{-s}}{s^2(s^2+3)}.$$

(c) To compute the inverse Laplace transform, we note that

$$\mathcal{L}^{-1}\left[\frac{2s}{s^2+3}\right] = 2\cos\sqrt{3}\,t.$$

Next, we note by partial fractions that

$$\frac{1}{s^2(s^2+3)} = \frac{1}{3}\left(\frac{1}{s^2} - \frac{1}{s^2+3}\right)$$

so

$$\mathcal{L}^{-1}\left[\frac{1-e^{-s}}{s^2(s^2+3)}\right] = \frac{1}{3}\left(t - \frac{1}{\sqrt{3}}\sin(\sqrt{3}t)\right) - \frac{1}{3}u_1(t)\left((t-1) + \frac{1}{\sqrt{3}}\sin(\sqrt{3}(t-1))\right).$$

Combining these two inverses, we obtain the solution

$$y(t) = 2\cos\sqrt{3}\,t + \frac{1}{3}\left(t - \frac{1}{\sqrt{3}}\sin(\sqrt{3}t)\right) - \frac{1}{3}u_1(t)\left((t-1) + \frac{1}{\sqrt{3}}\sin(\sqrt{3}(t-1))\right).$$

EXERCISES FOR SECTION 6.4

1. This is the $\frac{0}{0}$ case of L'Hôpital's Rule. Differentiating numerator and denominator with respect to Δt, we obtain

$$\frac{se^{s\Delta t} - (-s)e^{-s\Delta t}}{2},$$

which simplifies to

$$\frac{s(e^{s\Delta t} + e^{-s\Delta t})}{2}.$$

Since both $e^{s\Delta t}$ and $e^{-s\Delta t}$ tend to 1 as $\Delta t \to 0$, the desired limit is s.

3. Applying the Laplace transform to both sides, using the rules, and the fact that $\mathcal{L}[\delta_3] = e^{-3s}$, we get

$$s^2\mathcal{L}[y] - sy(0) - y'(0) + 2s\mathcal{L}[y] - 2y(0) + 5\mathcal{L}[y] = e^{-3s}.$$

Substituting the given initial conditions, we have

$$\mathcal{L}[y] = \frac{s+3}{s^2 + 2s + 5} + \frac{e^{-3s}}{s^2 + 2s + 5}.$$

Using the fact that $s^2 + 2s + 5 = (s+1)^2 + 4$, we obtain

$$\mathcal{L}[y] = \frac{s+1}{(s+1)^2 + 4} + \frac{2}{(s+1)^2 + 4} + \frac{1}{2}e^{-3s}\frac{2}{(s+1)^2 + 4}.$$

Therefore,

$$y(t) = e^{-t}\cos 2t + e^{-t}\sin 2t + \frac{1}{2}u_3(t)e^{-(t-3)}\sin(2(t-3)).$$

5. Applying Laplace transform to both sides, using the rules, and the fact that $\mathcal{L}[\delta_a] = e^{-as}$, we get

$$s^2\mathcal{L}[y] - sy(0) - y'(0) + 2s\mathcal{L}[y] - 2y(0) + 3\mathcal{L}[y] = e^{-s} - 3e^{-4s}.$$

Substituting the initial conditions gives us

$$\mathcal{L}[y] = \frac{e^{-s}}{s^2 + 2s + 3} - \frac{3e^{-4s}}{s^2 + 2s + 3}.$$

Now, using that $s^2 + 2s + 3 = (s+1)^2 + 2$, we have

$$\mathcal{L}[y] = \frac{1}{\sqrt{2}}e^{-s}\frac{\sqrt{2}}{(s+1)^2 + 2} + \frac{3}{\sqrt{2}}e^{-4s}\frac{\sqrt{2}}{(s+1)^2 + 2}.$$

So,

$$y(t) = \frac{1}{\sqrt{2}}u_1(t)e^{-(t-1)}\sin(\sqrt{2}(t-1)) - \frac{3}{\sqrt{2}}u_4(t)e^{-(t-4)}\sin(\sqrt{2}(t-4)).$$

7. **(a)** From the table

$$\mathcal{L}[\delta_a] = e^{-as}$$

$$s\mathcal{L}[u_a] - u_a(0) = s\frac{e^{-as}}{s} - 0 = e^{-as}.$$

(b) The formula for the Laplace transform of a derivative is

$$\mathcal{L}\left[\frac{dy}{dt}\right] = s\mathcal{L}[y] - y(0)$$

and this is exactly the relationship between the Laplace transforms of $u_a(t)$ and $\delta_a(t)$. Hence, it is tempting to think of the Dirac delta function as the derivative of the Heaviside function.

(c) We can think of the Heaviside function $u_a(t)$ as a limit of piecewise linear functions equal to zero for t less than $a - \Delta t$, equal to one for t greater than $a + \Delta t$ and a straight line for t between $a - \Delta t$ and $a + \Delta t$. The derivative of this function is precisely the function $g_{\Delta t}$ used to define the Dirac delta function. This is still just an informal relationship until we specify in what sense we are taking the limit.

9. **(a)** To compute the Laplace transform of the sum on the right-hand side we can either work "by hand" computing the Laplace transform of each term ($\mathcal{L}[\delta_n] = e^{-ns}$) and summing the resulting geometric series or use Exercise 16 in Section 6.2. In either case

$$\mathcal{L}\left[\sum_{n=1}^{\infty} \delta_n(t)\right] = \frac{e^{-s}}{1 - e^{-s}} = \frac{1}{e^s - 1}.$$

For our purposes it is actually better to leave the Laplace transform of the right-hand side as

$$\mathcal{L}\left[\sum_{n=1}^{\infty} \delta_n(t)\right] = \sum_{n=1}^{\infty} e^{-ns}.$$

Hence, applying Laplace transforms to both sides and simplifying gives

$$\mathcal{L}[y] = \frac{1}{s^2 + 2}\sum_{n=1}^{\infty} e^{-ns}.$$

(b) Now taking the inverse Laplace transform (term wise in the sum) gives

$$y(t) = \frac{1}{\sqrt{2}}\sum_{n=1}^{\infty} u_n(t)\sin(\sqrt{2}(t - n)).$$

(c) The period of the forcing is different from the natural period of the unforced oscillator. Hence the solution oscillates, with rising and falling period. There is a discontinuity $t = n$ for any integer n. Looking at the formula for the solution in part (b), we see that some of the signs of the terms will not be consistent. For $t < N$, only the terms with $n < N$ are nonzero and for a given t value some of these are positive and some are negative.

EXERCISES FOR SECTION 6.5

1. Using the definition of the convolution with f and g, we see that

$$(f * g)(t) = \int_0^t 1 \cdot e^{-u} \, du$$

$$= \int_0^t e^{-u} \, du$$

$$= -e^{-u} \Big|_0^t$$

$$= 1 - e^{-t}.$$

Checking the convolution property $(\mathcal{L}[f * g] = \mathcal{L}[f] \cdot \mathcal{L}[g])$ for Laplace transforms, we have

$$\mathcal{L}[f] = \frac{1}{s}, \quad \mathcal{L}[g] = \frac{1}{s+1},$$

and

$$\mathcal{L}[f * g] = \frac{1}{s} - \frac{1}{s+1}$$

$$= \frac{s+1-s}{s(s+1)}$$

$$= \frac{1}{s(s+1)}.$$

So, $\mathcal{L}[f] \cdot \mathcal{L}[g] = \mathcal{L}[f * g]$.

3. Using the definition of the convolution with f and g, we see that

$$(f * g)(t) = \int_0^t \cos(t - v)u_2(v) \, dv.$$

(We're using v as the integrating variable instead of u so as not to confuse it with the Heaviside function.) First, notice that if $0 < v < 2$, then $u_2(v) = 0$. Thus, if $t < 2$, then the function $u_2(v)$ is always 0, which means the integral is 0. Now, if $t \geq 2$, then

$$\int_0^t \cos(t - v)u_2(v) \, dv = \int_0^2 \cos(t - v) \, 0 \, dv + \int_2^t \cos(t - v) \, 1 \, dv$$

$$= \int_2^t \cos(t - v) \, dv.$$

So,

$$\int_0^t \cos(t - v)u_2(v) \, dv = \begin{cases} 0, & \text{if } t < 2, \\ \int_2^t \cos(t - v) \, dv, & \text{if } t \geq 2. \end{cases}$$

Evaluating the second integral, we get

$$\int_2^t \cos(t - v)\, dv = -\sin(t - v)\Big|_2^t = \sin(t - 2).$$

We have a function that is 0 for $t < 2$ and equal to $\sin(t - 2)$ for $t \geq 2$, so our function is

$$u_2(t)\sin(t - 2).$$

Checking the convolution property ($\mathcal{L}[f * g] = \mathcal{L}[f] \cdot \mathcal{L}[g]$) for Laplace transforms, we have

$$\mathcal{L}[f] = \frac{s}{s^2 + 1}, \quad \mathcal{L}[g] = \frac{e^{-2s}}{s},$$

and

$$\mathcal{L}[f * g] = \frac{e^{-2s}}{s^2 + 1}.$$

Hence, $\mathcal{L}[f] \cdot \mathcal{L}[g] = \mathcal{L}[f * g]$.

5. Using the definition of the convolution with f and g, we see that

$$(f * g)(t) = \int_0^t 3\sin(t - u)\cos(2u)\, du.$$

We will use four trigonometric identities to evaluate this integral:

$$\sin(t - u) = \sin t \cos u - \cos t \sin u$$

$$\sin(mt)\sin(nt) = \tfrac{1}{2}[\cos((m - n)t) - \cos((m + n)t)]$$

$$\cos(mt)\cos(nt) = \tfrac{1}{2}[\cos((m + n)t) + \cos((m - n)t)]$$

$$\sin(mt)\cos(nt) = \tfrac{1}{2}[\sin((m + n)t) + \sin((m - n)t)].$$

So

$$\int_0^t 3\sin(t - u)\cos(2u)\, du$$

$$= \int_0^t [3\cos 2u \cos u \sin t - \cos 2u \sin u \cos t]\, du$$

$$= \int_0^t \left[\tfrac{3}{2}(\cos 3u + \cos u)\sin t - \tfrac{1}{2}(\sin 3u - \sin u)\cos t\right] du$$

$$= \sin t \left[\tfrac{1}{2}\sin 3u + \tfrac{3}{2}\sin u\right]_0^t + \cos t \left[\tfrac{1}{2}\cos 3u - \tfrac{3}{2}\cos u\right]_0^t$$

$$= \sin t \left(\tfrac{1}{2}\sin 3t + \tfrac{3}{2}\sin t\right) + \cos t \left(\tfrac{1}{2}\cos 3t - \tfrac{3}{2}\cos t + 1\right)$$

$$= \tfrac{1}{2}\sin 3t \sin t + \tfrac{3}{2}\sin^2 t + \tfrac{1}{2}\cos 3t \cos t - \tfrac{3}{2}\cos^2 t + \cos t$$

$$= \tfrac{1}{4}(\cos 2t - \cos 4t) + \tfrac{3}{2}\sin^2 t + \tfrac{1}{4}(\cos 4t + \cos 2t) - \tfrac{3}{2}\cos^2 t + \cos t$$

$$= \tfrac{1}{2}\cos 2t + \tfrac{3}{2}\left(\sin^2 t - \cos^2 t\right) + \cos t$$

$$= \tfrac{1}{2}\cos 2t - \tfrac{3}{2}\cos 2t + \cos t$$

$$= \cos t - \cos 2t,$$

which is the same answer obtained in the text using the technique of Laplace transforms.

7. Taking Laplace transform of both sides of the equation and solving for $\mathcal{L}[\zeta]$ (see page 589), we obtain

$$\mathcal{L}[\zeta] = \frac{1}{s^2 + ps + q}.$$

Hence, if we let

$$z(s) = s^2 + ps + q,$$

we have that $z(0) = 5$ and $z(2) = 17$. Now $z(0) = 5$ implies $q = 5$. Using $z(2) = 17 = 2^2 + 2p + 5$, we see that $p = 4$.

9. (a) Since ζ solves the initial-value problem above, we know that

$$\frac{d\zeta^2}{dt^2} + p\frac{d\zeta}{dt} + q\zeta = \delta_0(t), \quad \zeta(0) = \zeta'(0) = 0^-.$$

Taking Laplace transforms of both sides, and substituting initial conditions gives us

$$s^2 \mathcal{L}[\zeta] + ps\mathcal{L}[\zeta] + q\mathcal{L}[\zeta] = 1,$$

which yields

$$\mathcal{L}[\zeta] = \frac{1}{s^2 + ps + q}.$$

Now, taking Laplace transforms of both sides of

$$\frac{d^2 y}{dt^2} + p\frac{dy}{dt} + qy = 0, \quad y(0) = a, \, y'(0) = 0$$

gives us

$$s^2 \mathcal{L}[y] - sa + ps\mathcal{L}[y] - pa + q\mathcal{L}[y] = 0.$$

Solving for $\mathcal{L}[y]$ gives

$$\mathcal{L}[y] = \frac{a(s + p)}{s^2 + ps + q},$$

so

$$\mathcal{L}[y] = a(s + p)\mathcal{L}[\zeta].$$

(b) Taking Laplace transforms of both sides of

$$\frac{d^2 y}{dt^2} + p\frac{dy}{dt} + qy = 0, \quad y(0) = 0, \, y'(0) = b$$

gives us

$$s^2 \mathcal{L}[y] - b + ps\mathcal{L}[y] + q\mathcal{L}[y] = 0.$$

Solving for $\mathcal{L}[y]$ gives

$$\mathcal{L}[y] = \frac{b}{s^2 + ps + q},$$

so

$$\mathcal{L}[y] = b\mathcal{L}[\zeta].$$

(c) Taking Laplace transforms of both sides of

$$\frac{d^2y}{dt^2} + p\frac{dy}{dt} + qy = f(t), \quad y(0) = a, \, y'(0) = b$$

gives us

$$s^2 \mathcal{L}[y] - sa - b + ps\mathcal{L}[y] - pa + q\mathcal{L}[y] = \mathcal{L}[f].$$

Solving for $\mathcal{L}[y]$ gives

$$\mathcal{L}[y] = \frac{\mathcal{L}[f] + a(s + p) + b}{s^2 + ps + q},$$

so

$$\mathcal{L}[y] = (\mathcal{L}[f] + a(s + p) + b)\,\mathcal{L}[\zeta].$$

11. **(a)** We know that

$$\frac{d^2y_1}{dt^2} + p\frac{dy_1}{dt} + qy_1 = f_1(t).$$

Taking Laplace transforms of both sides and solving for $\mathcal{L}[y_1]$ gives

$$\mathcal{L}[y_1] = \frac{\mathcal{L}[f_1]}{s^2 + ps + q}.$$

(b) As above, we know that

$$\frac{d^2y_2}{dt^2} + p\frac{dy_2}{dt} + qy_2 = f_2(t),$$

so

$$\mathcal{L}[y_2] = \frac{\mathcal{L}[f_2]}{s^2 + ps + q}.$$

Now we see that

$$\frac{\mathcal{L}[f_1]}{\mathcal{L}[y_1]} = s^2 + ps + q,$$

and

$$\frac{\mathcal{L}[f_2]}{\mathcal{L}[y_2]} = s^2 + ps + q,$$

so

$$\frac{\mathcal{L}[f_1]}{\mathcal{L}[y_1]} = \frac{\mathcal{L}[f_2]}{\mathcal{L}[y_2]}.$$

(c) Solving for $\mathcal{L}[y_2]$ gives us

$$\mathcal{L}[y_2] = \mathcal{L}[f_2]\frac{\mathcal{L}[y_1]}{\mathcal{L}[f_1]}.$$

EXERCISES FOR SECTION 6.6

1. **(a)** Taking Laplace transform of both sides and substituting the initial conditions yields

$$s^2\mathcal{L}[y] - 2s + 2 + 2s\mathcal{L}[y] - 4 + 2\mathcal{L}[y] = \frac{4}{(s+2)^2 + 16}.$$

Hence,

$$\mathcal{L}[y] = \frac{2s+2}{s^2 + 2s + 2} + \frac{4}{(s^2 + 4s + 20)(s^2 + 2s + 2)}.$$

(b) The poles are the roots of $s^2 + 2s + 2$ and $s^2 + 4s + 20$, or $-1 \pm i$ and $-2 \pm 4i$.

(c) Since all poles have negative real parts, the solution tends to zero at an exponential rate. The real part closest to 0 is -1 so solutions tend to zero at a rate of e^{-t}. Since the poles are complex, the solutions oscillate. The oscillations with period $2\pi/4 = \pi/2$ decay at the rate e^{-2t} while the oscillations with period 2π decay at the rate e^{-t}.

3. **(a)** Taking the Laplace transform of both sides and plugging in the initial conditions gives

$$s^2\mathcal{L}[y] + s\mathcal{L}[y] + 8\mathcal{L}[y] = \mathcal{L}[\cos(t-4)] - \mathcal{L}[u_4(t)\cos(t-4)].$$

To take the Laplace transform of the first term on the right we recall that

$$\cos(t-4) = \cos(4)\cos(t) + \sin(4)\sin(t),$$

so

$$(s^2 + s + 8)\mathcal{L}[y] = \cos(4)\frac{s}{s^2+1} + \sin(4)\frac{1}{s^2+1} - \frac{se^{-4s}}{s^2+1}.$$

Therefore,

$$\mathcal{L}[y] = \frac{\cos(4)s + \sin(4) - se^{-4s}}{(s^2 + s + 8)(s^2 + 1)}.$$

(b) The poles are given by $s^2 + 1 = 0$ and $s^2 + s + 8 = 0$ or $s = \pm i$ and $s = -1/2 \pm i\sqrt{31}/2$.

(c) Note that the forcing "turns off" at time $t = 4$. Hence up until time $t = 4$ there is a forced response with period 2π. Since $t = 4$ is only about half a period for the forced response, this is not very significant. The natural response is an oscillation with period $4\pi/\sqrt{31}$ which decays like $e^{-t/2}$, and this is the long-term behavior of the solution.

5. **(a)** Computing the Laplace transform of both sides of the equation and using the initial conditions gives

$$s^2\mathcal{L}[y] - s - 1 + 16\mathcal{L}[y] = 0,$$

so

$$\mathcal{L}[y] = \frac{s+1}{s^2+16}.$$

(b) The poles are the solutions of $s^2 + 16 = 0$ or $s = \pm 4i$.

(c) If we find the inverse Laplace transform, we see that

$$y(t) = \cos(4t) + (1/4)\sin(4t).$$

So a reasonable conjecture would be that when the poles of the Laplace transform are on the imaginary axis, the solution is periodic and does not decay.

7. (a) As usual, we take the Laplace transform of both sides of the equation and substitute the initial conditions to get

$$s^2 \mathcal{L}[y] - s - 2 + 2s \mathcal{L}[y] - 2 + \mathcal{L}[y] = 0.$$

Hence,

$$\mathcal{L}[y] = \frac{s+4}{s^2 + 2s + 1}.$$

(b) The poles are the roots of $s^2 + 2s + 1 = (s+1)^2$. Hence, there is a "double pole" at $s = -1$.

(c) For homogeneous second-order equations, double poles play the same role as double eigenvalues. In the case of a double pole on the real axis, the solution is a critically damped oscillator.

9. (a) As usual, we compute

$$s^2 \mathcal{L}[y] - sy(0) - y'(0) + 20s \mathcal{L}[y] - 20y(0) + 200\mathcal{L}[y] = \mathcal{L}[w(t)]$$

and using the initial conditions and Exercise 17 of Section 6.2, we have that

$$(s^2 + 20s + 200)\mathcal{L}[y] - s - 20 = \frac{1 - e^{-s}}{s(1 + e^{-s})}.$$

Hence,

$$\mathcal{L}[y] = \frac{s + 20}{s^2 + 20s + 200} + \frac{1 - e^{-s}}{s(1 + e^{-s})(s^2 + 20s + 200)}.$$

(b) The poles are the roots of $s^2 + 20s + 200 = 0$ ($s = -10 \pm 10i$) and the zeros of $s(1 + e^{-s}) = 0$. One zero is $s = 0$, and using Euler's formula, we obtain the other zeros $s = (2n + 1)i\pi$ for $n = 0, \pm 1, \pm 2, \dots$.

(c) The natural response corresponds to the poles $s = -10 \pm 10i$, so it decays like e^{-10t} (very rapidly), while oscillating with a period of $\pi/5$. The remainder of the poles correspond to the forcing and indicate a forced response, which is an oscillation between $\pm 1/200$. Because the forcing is discontinuous, solutions settle (quickly) toward $1/200$ for $0 < t < 1$. When the forcing switches, they tend (quickly) to $-1/200$. (The function $y(t) = 1/200$ is a particular solution of the equation

$$\frac{d^2 y}{dt^2} + 20\frac{dy}{dt} + 200y = 1,$$

and $y(t) = -1/200$ is a solution for the same equation if the forcing is -1 rather than 1.)

CHAPTER 7

Numerical Methods

EXERCISES FOR SECTION 7.1

1. **(a)** The differential equation is separable. Therefore, one way to obtain the solution to the initial-value problem is to integrate

$$\int y^{-2}\,dy = \int -2t\,dt.$$

We obtain

$$\frac{y^{-1}}{-1} = -t^2 + c$$

$$y^{-1} = t^2 + k$$

$$y = \frac{1}{t^2 + k}.$$

We determine the value of k using the initial condition $y(0) = 1$. Hence, $k = 1$, and the solution to the given initial-value problem is $y(t) = 1/(t^2 + 1)$.

(b) To calculate y_{20}, we must apply Euler's method 20 times. Table 7.1 contains the results of a number of intermediate calculations.

(c) The total error e_{20} is the difference between the actual value $y(2) = 0.2$ and the approximate value $y_{20} = 0.193342$. Therefore, $e_{20} = 0.0066581$.

(d) Table 7.2 contains the results of Euler's method and the corresponding total errors for $n = 1000$, $2000, \ldots, 6000$.

Table 7.1
Results of Euler's method

k	t_k	y_k	k	t_k	y_k
0	0	1.0	10	1.0	0.503642
1	0.1	1.0	⋮	⋮	⋮
2	0.2	0.98	19	1.9	0.210119
3	0.3	0.941584	20	2.0	0.193342
4	0.4	0.888389			
5	0.5	0.82525			

Table 7.2
Results of Euler's method and the corresponding total errors

n	y_n	e_n
1000	0.199874	0.000125848
2000	0.199937	0.0000628901
3000	0.199958	0.0000419192
4000	0.199969	0.0000314366
5000	0.199975	0.0000251479
6000	0.199979	0.0000209558

(e) Table 7.3 gives values of e_n for some n that are intermediate to the ones above in case that you want to double check the ones you have computed. Also, the graph of e_n as a function of n for $100 \leq n \leq 6000$ is given.

Table 7.3
Selected total errors

n	e_n
100	0.00127095
200	0.000631985
1400	0.0000898639
3700	0.0000339862
5600	0.000022453

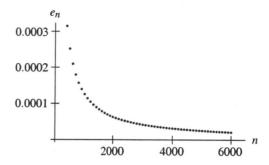

(f) Our computer math system fits the data to the function $0.126731/n$. The following figure includes both the data and the graph of this function.

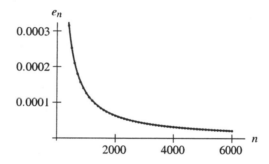

3. (a) The differential equation is both separable and linear. Therefore, one way to obtain the solution to the initial-value problem is to integrate

$$\int \frac{1}{y}\,dy = \int t\,dt.$$

We obtain

$$\ln|y| = \frac{t^2}{2} + c$$

$$y = ke^{t^2/2}.$$

We determine the value of k using the initial condition $y(0) = 1$. Hence, $k = 1$, and the solution to the given initial-value problem is $y(t) = e^{t^2/2}$.

(b) To calculate y_{20}, we must apply Euler's method 20 times. Table 7.4 contains the results of a number of intermediate calculations.

(c) The total error e_{20} is the difference between the actual value $y(\sqrt{2}) = e$ and the approximate value $y_{20} = 2.51066$. Therefore, $e_{20} = 0.20762$.

(d) Table 7.5 contains the results of Euler's method and the corresponding total errors for $n = 1000$, $2000, \ldots, 6000$.

Table 7.4
Results of Euler's method

k	t_k	y_k	k	t_k	y_k
0	0	1	10	0.707107	1.24797
1	0.0707107	1	⋮	⋮	⋮
2	0.141421	1.005	19	1.3435	2.29284
3	0.212132	1.01505	20	1.41421	2.51066
4	0.282843	1.03028			
5	0.353553	1.05088			

Table 7.5
Results of Euler's method and the corresponding total errors

n	y_n	e_n
1000	2.71376	0.00452218
2000	2.71602	0.00226316
3000	2.71677	0.00150923
4000	2.71715	0.0011321
5000	2.71738	0.000905762
6000	2.71753	0.000754848

(e) Table 7.6 gives values of e_n for some n that are intermediate to the ones above in case that you want to double check the ones you have computed. Also, the graph of e_n as a function of n for $100 \leq n \leq 6000$ is given.

Table 7.6
Selected total errors

n	e_n
100	0.0444901
200	0.0224467
1400	0.00323182
3700	0.00122384
5600	0.000808748

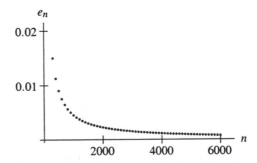

(f) Our computer math system fits the data to the function $4.47023/n$. The following figure includes both the data and the graph of this function.

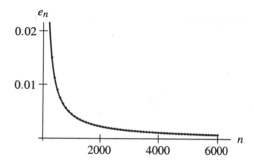

5. **(a)** The differential equation is linear. Therefore, we can use integrating factors to find an analytic solution (see Section 1.8). In this case, the integrating factor is $\mu(t) = e^{-3t}$. We rewrite the differential equation in the form

$$\frac{dy}{dt} - 3y = 1 - t$$

and multiply both sides by the integrating factor e^{-3t}. We obtain

$$e^{-3t}\frac{dy}{dt} - 3e^{-3t}y = e^{-3t}(1 - t),$$

which is equivalent to

$$\frac{d\left(ye^{-3t}\right)}{dt} = e^{-3t}(1 - t).$$

Integrating both sides with respect to t (using integration by parts on the right-hand side) yields

$$ye^{-3t} = e^{-3t}\left(\frac{t}{3} - \frac{2}{9}\right) + c$$

where c is a constant of integration. Multiplying through by e^{3t} gives

$$y = \left(\frac{t}{3} - \frac{2}{9}\right) + ce^{3t}.$$

This result is the general solution to this linear differential equation.

We determine the value of c using the initial condition $y(0) = 1$. Hence, $c = 11/9$, and the solution to the given initial-value problem is

$$y(t) = \frac{11e^{3t}}{9} + \frac{t}{3} - \frac{2}{9}.$$

(b) To calculate y_{20}, we must apply Euler's method 20 times. Table 7.7 contains the results of a number of intermediate calculations.

(c) The total error e_{20} is the difference between the actual value $y(1) = (11e^3 + 1)/9$ and the approximate value $y_{20} = 20.1147$. Therefore, $e_{20} = 4.54544$.

(d) Table 7.8 contains the results of Euler's method and the corresponding total errors for $n = 1000$, $2000, \ldots, 6000$.

Table 7.7
Results of Euler's method

k	t_k	y_k	k	t_k	y_k
0	0	1.	10	0.5	4.88902
1	0.05	1.2	⋮	⋮	⋮
2	0.1	1.4275	19	0.95	17.4888
3	0.15	1.68663	20	1.	20.1147
4	0.2	1.98212			
5	0.25	2.31944			

Table 7.8
Results of Euler's method and the corresponding total errors

n	y_n	e_n
1000	24.5501	0.110003
2000	24.605	0.0551181
3000	24.6233	0.0367714
4000	24.6325	0.0275883
5000	24.638	0.0220753
6000	24.6417	0.0183987

(e) Table 7.3 gives values of e_n for some n that are intermediate to the ones above in case that you want to double check the ones you have computed. Also, the graph of e_n as a function of n for $100 \leq n \leq 6000$ is given.

Table 7.9
Selected total errors

n	e_n
100	1.05955
200	0.540843
1400	0.0786686
3700	0.0298226
5600	0.0197119

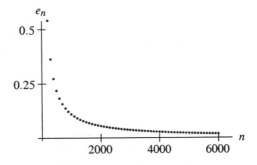

(f) Our computer math system fits the data to the function $107.125/n$. The following figure includes both the data and the graph of this function.

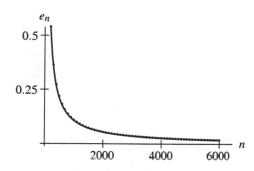

7. **(a)** The partial derivative $\partial f/\partial y$ of $f(t, y) = -2ty^2$ is $-4ty$. Thus, $M_2 = -4ty$.
The partial derivative $\partial f/\partial t$ of $f(t, y) = -2ty^2$ is $-2y^2$. Therefore,

$$M_1 = \frac{\partial f}{\partial t} + \frac{\partial f}{\partial y} f(t, y) = -2y^2 + (-4ty)(-2ty^2) = -2y^2 + 8t^2y^3.$$

(b) Since $f(t_0, y_0) = 0$, the result of the first step of Euler's method is the point $(t_1, y_1) = (0.02, 1)$. Consequently, to estimate the error, we compute the quantity M_1 at the point $(0.01, 1)$. We obtain $M_1 \approx -1.9992$.

Once we have M_1 at this point, we can estimate the error e_1 (which is the same as the truncation error) by computing $|M_1(\Delta t)^2/2|$. In this case, we obtain 0.00039984.

To calculate the actual error, we can compare the result y_1 of Euler's method with the value of the solution $y(0.2)$. Since we know that the solution is $1/(1 + t^2)$, we have $y(0.2) = 0.9996$, and thus the actual error is 0.00039984. For this computation, note that the estimated error and the actual error essentially agree. (In order to see a difference in these two quantities, we had to do the calculations to 11 decimal places.)

(c) The second point (t_2, y_2) obtained from Euler's method is the point $(0.04, 0.9992)$. For the second step, the estimated error is no longer simply the truncation error, so we must compute both M_1 and M_2. Evaluating M_1 at the point $(0.03, 0.9996)$, we obtain $M_1 = -1.99121$. Evaluating M_2 at the point $(0.02, 0.9996)$, we obtain $M_2 = -0.079968$.

Now to estimate the error in the second step, we use the approximation

$$e_k \approx (1 + M_2\Delta t) e_{k-1} + M_1 \frac{(\Delta t)^2}{2}$$

and obtain $e_2 \approx 0.000797442$.

To compare this estimate to the actual error, we compute the value of the solution $y(0.04) = 0.998403$. Since $y_2 = 0.9992$, we see that the actual error e_2 is 0.000797444. Note that our estimate of e_2 and the true value of e_2 are very close.

(d) Table 7.10 gives the values of e_k and our estimates of e_k for $k = 10, 20, 30, \ldots, 100$ in case that you want to double check your computations.

Table 7.10
Selected total errors

k	t_k	e_k	estimated e_k
10	0.2	-0.00354477	-0.00354363
20	0.4	-0.00487796	-0.00486646
30	0.6	-0.00399684	-0.00396879
40	0.8	-0.00226733	-0.00223171
50	1.	-0.000714495	-0.000683793
60	1.2	0.000335713	0.000355439
70	1.4	0.000929043	0.000937796
80	1.6	0.00120617	0.00120657
90	1.8	0.00129161	0.0012866
100	2.	0.00127095	0.00126289

(e) Compare your plots with the Figures 7.5 and 7.6 in Section 7.1.

9. (a) The partial derivative $\partial f / \partial y$ of $f(t, y) = \sin ty$ is $t \cos ty$. Thus, $M_2 = t \cos ty$.
The partial derivative $\partial f / \partial t$ of $f(t, y) = \sin ty$ is $y \cos ty$. Therefore,

$$M_1 = \frac{\partial f}{\partial t} + \frac{\partial f}{\partial y} f(t, y) = (y + t \sin ty) \cos ty.$$

(b) The result of the first step of Euler's method is the point $(t_1, y_1) = (0.03, 3.0)$. Consequently, to estimate the error, we compute the quantity M_1 at the point $(0.015, 3)$. We obtain $M_1 \approx 2.99764$.
Once we have M_1 at this point, we can estimate the error e_1 (which is the same as the truncation error) by computing $|M_1 (\Delta t)^2 / 2|$. In this case, we obtain 0.00134894.
We cannot compare our estimate to the true error since we do not know how to calculate the true error for this differential equation. (We cannot find a closed-form solution.)

(c) The second point (t_2, y_2) obtained from Euler's method is the point $(0.06, 3.0027)$. For the second step, the estimated error is no longer simply the truncation error, so we must compute both M_1 and M_2. Evaluating M_1 at the point $(0.045, 3.00135)$, we obtain $M_1 = 2.98002$. Evaluating M_2 at the point $(0.03, 3.00135)$, we obtain $M_2 = 0.0298785$.
Now to estimate the error in the second step, we use the approximation

$$e_k \approx (1 + M_2 \Delta t) e_{k-1} + M_1 \frac{(\Delta t)^2}{2}$$

and obtain $e_2 \approx 0.00269115$.

(d) The following table gives the values of our estimates of e_k for $k = 10, 20, 30, \ldots, 100$ in case that you want to double check your computations. We also plot our estimates of the error as a function of k.

Table 7.11
Selected error estimates

k	t_k	estimated e_k
10	0.3	0.0123521
20	0.6	0.0138193
30	0.9	0.00135629
40	1.2	0.0109331
50	1.5	0.0121639
60	1.8	0.0119345
70	2.1	0.0129518
80	2.4	0.0164855
90	2.7	0.0240115
100	3.	0.0343504

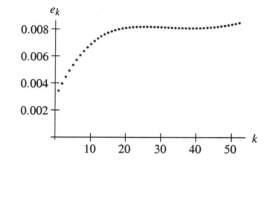

11. **(a)** The argument that justifies the inequality

$$e_1 \leq M_1 \frac{(\Delta t)^2}{2}$$

is given on pages 611 and 612. In particular, the truncation error in the first step is given by Taylor's Theorem.

(b) The total error e_2 involved in the second step is discussed on pages 613 and 614. On the right-hand side of the inequality

$$e_2 \leq e_1 + M_2 e_1 \Delta t + M_1 \frac{(\Delta t)^2}{2},$$

the first term is the error involved in the previous step. The second term measures the contribution to the error that arises from evaluating the right-hand side of the differential equation at the point (t_1, y_1) rather than at $(t_1, y(t_1))$. The third term measures the truncation error associated this step of the approximation (see Figure 7.4).

(c) The analysis of the error e_{k+1} is essentially identical to the analysis of the error e_2. That is,

$$e_{k+1} = |y(t_{k+1}) - y_{k+1}|$$

where $y(t_{k+1})$ is the actual value of the function and y_{k+1} is the value given by Euler's method. In other words,

$$y_{k+1} = y_k + f(t_k, y_k)\, \Delta t.$$

Applying Taylor's Theorem to $y(t)$ at the point $(t_k, y(t_k))$, we have

$$y(t_{k+1}) = y(t_k) + f(t_k, y(t_k)) + y''(\xi_k) \frac{(\Delta t)^2}{2}.$$

Thus,

$$e_{k+1} = |y(t_{k+1}) - y_{k+1}|$$

$$\leq |y(t_k) + f(t_k, y(t_k)) + y''(\xi_k)\frac{(\Delta t)^2}{2} - (y_k + f(t_k, y_k)\,\Delta t)|$$

$$\leq |y(t_k) - y_k| + |f(t_k, y(t_k)) - f(t_k, y_k)|\,\Delta t + |y''(\xi_k)|\frac{(\Delta t)^2}{2}$$

$$\leq e_k + |f(t_k, y(t_k)) - f(t_k, y_k)|\,\Delta t + |y''(\xi_k)|\frac{(\Delta t)^2}{2}.$$

The term $|f(t_k, y(t_k)) - f(t_k, y_k)|$ is bounded by the product of M_2 and e_k, and the third term (the truncation error) is bounded by $M_1(\Delta t)^2/2$. Hence, we have

$$e_{k+1} \leq e_k + (M_2)(e_k)(\Delta t) + M_1\frac{(\Delta t)^2}{2}$$

$$= (1 + M_2\Delta t)e_k + M_1\frac{(\Delta t)^2}{2}.$$

(d) We know that $e_1 \leq K_2$ because e_1 is the same as the truncation error. Using the result of part (b) (or part (c) with $k = 1$), we know that

$$e_2 \leq (1 + M_2\Delta t)e_1 + M_1\frac{(\Delta t)^2}{2} = (K_1)e_1 + K_2.$$

However, given that $e_1 \leq K_2$, we have

$$e_2 \leq (K_1 + 1)K_2.$$

Finally, from part (c), we know that

$$e_3 \leq (K_1)e_2 + K_2 \leq (K_1)(K_1 + 1)K_2 + K_2 = (K_1^2 + K_1 + 1)K_2.$$

(e) We can verify this assertion by induction. In fact, we can use the result of part (d) as the first step. Then we assume the inductive hypothesis that

$$e_{n-1} \leq \left(K_1^{n-2} + K_1^{n-3} + \cdots + K_1 + 1\right)K_2.$$

Since part (c) says that $e_n \leq (K_1)(e_{n-1}) + K_2$, we can use the inductive hypothesis to obtain

$$e_n \leq (K_1)\left(K_1^{n-2} + K_1^{n-3} + \cdots + K_1 + 1\right)K_2 + K_2$$

$$= \left(K_1^{n-1} + K_1^{n-2} + \cdots + K_1 + 1\right)K_2.$$

(f) We can verify the formula

$$K_1^{n-1} + K_1^{n-2} + \cdots + K_1 + 1 = \frac{K_1^n - 1}{K_1 - 1}$$

using induction. However, it is probably easier to see why it holds if we multiply both sides by the factor $(K_1 - 1)$ and cancel on the left-hand side.

Applying this formula to the result of part (e), we get

$$e_n \leq \left(\frac{K_1^n - 1}{K_1 - 1} \right) K_2.$$

(g) Since $K_1 = 1 + M_2 \Delta t$ and $K_2 = M_2 (\Delta t)^2 / 2$, we have

$$e_n \leq \left(\frac{(1 + M_2 \Delta t)^n - 1}{M_2 \Delta t} \right) \left(M_1 \frac{(\Delta t)^2}{2} \right)$$

$$= \frac{M_1}{2M_2} \left((1 + M_2 \Delta t)^n - 1 \right) \Delta t.$$

(h) If we let $\alpha = M_2 \Delta t$ in Exercise 10, then

$$\left((1 + M_2 \Delta t)^n - 1 \right) \leq \left(\left(e^{(M_2 \Delta t)} \right)^n - 1 \right)$$

$$= \left(e^{(M_2 \Delta t)n} - 1 \right).$$

Consequently, we can conclude that

$$e_n \leq \frac{M_1}{2M_2} \left(e^{(M_2 \Delta t)n} - 1 \right) \Delta t.$$

(i) Since $\Delta t = (t_n - t_0)/n$, the product $(M_2 \Delta t)n$ is equal to $M_2 (t_n - t_0)$. Therefore, we can conclude that

$$e_n \leq \frac{M_1}{2M_2} \left(e^{M_2(t_n - t_0)} - 1 \right) \Delta t.$$

(j) Note that the quantities M_1 and M_2 are determined by the right-hand side of the differential equation and the rectangle R in the ty-plane. Also, the quantity $(t_n - t_0)$ is precisely the length of the interval over which we are approximating the solution. Therefore, all of the terms in the expression

$$\frac{M_1}{2M_2} \left(e^{M_2(t_n - t_0)} - 1 \right)$$

do not depend on the number of steps involved in the application of Euler's method. In other words, when considering the effectiveness of Euler's method, we can treat this expression as a constant C determined only by the right-hand side of the differential equation and the rectangle R under consideration.

This expression includes all of the right-hand side of part (i) with the exception of the Δt factor. Consequently, we can consider this long and involved derivation as a justification of the simple inequality

$$e_n \leq C \cdot \Delta t.$$

(k) This estimate is a rigorous one. In other words, given the hypotheses stated at the beginning of the exercise, we can be certain that the error is indeed bounded by the quantities specified. The "estimates" calculated in Exercises 7–9 are not as certain. The logic that justifies their calculation is valid, but we cannot be certain that these quantities always give us a true indication of the accuracy of our approximations.

EXERCISES FOR SECTION 7.2

1. Table 7.12 includes the approximate values y_k obtained using improved Euler's method. In addition to the results of improved Euler's method, we graph the results of Euler's method and the results obtained when we used a built-in numerical solver.

Table 7.12
Results of improved Euler's method

k	t_k	y_k
0	0.0	3.0000
1	0.5	8.2500
2	1.0	21.3750
3	1.5	54.1875
4	2.0	136.2187

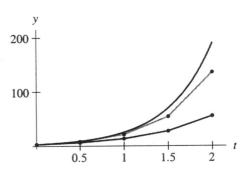

3. Table 7.13 includes the approximate values y_k obtained using improved Euler's method. In addition to the results of improved Euler's method, we graph the results of Euler's method and the results obtained when we used a built-in numerical solver. Note that the actual solution seems to blow up near $t = 1$, and Euler's method does not do a good job of predicting this. Improved Euler's method does a better job of suggesting that something unusual is happening with this solution.

Table 7.13
Results of improved Euler's method

k	t_k	y_k
0	0.0	2.000
1	0.5	2.813
2	1.0	6.618
3	1.5	129.001
4	2.0	17310268.856

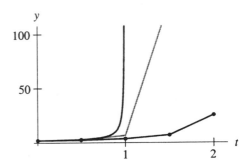

5. Table 7.14 includes the approximate values w_k obtained using improved Euler's method. At $t = 3$, we start to wonder about the quality of our results. A simple qualitative analysis of the differential equation indicates that $w = -1$ and $w = 3$ are equilibrium solutions and that solutions with initial conditions above 3 approach $w = 3$ as $t \to \infty$. Therefore, we know that the results of improved Euler's method in this case are useless. Compare the results obtained here with the results obtained in Exercise 6.

In addition to the results of improved Euler's method, we graph the results of Euler's method and the results obtained when we used a built-in numerical solver. Note that Euler's method yields an approximate solution that heads for the wrong equilibrium solution.

Table 7.14

Results of improved Euler's method

k	t_k	y_k
0	0.0	4.00
1	1.0	1.50
2	2.0	-3.66
3	3.0	-259.96

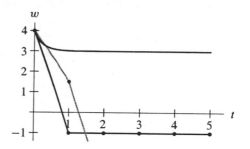

7. Table 7.15 includes the approximate values y_k obtained using improved Euler's method. In addition to the results of improved Euler's method, we graph the results of Euler's method and the results obtained when we used a built-in numerical solver.

Table 7.15

Results of improved Euler's method

k	t_k	y_k
0	0.	2.
1	0.5	3.13301
2	1.	4.01452
3	1.5	4.80396
4	2.	5.54124

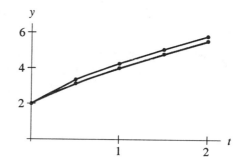

9. **(a)** The differential equation is both separable and linear. Therefore, one way to obtain the solution to the initial-value problem is to integrate

$$\int \frac{1}{1-y}\,dy = \int 1\,dt.$$

We obtain

$$-\ln|1 - y| = t + c,$$

which yields

$$y = 1 - ke^{-t}.$$

We determine the value of k using the initial condition $y(0) \doteq 0$. Hence, $k = 1$, and the solution to the given initial-value problem is $y(t) = 1 - e^{-t}$.

(b) Table 7.16 contains the steps involved in applying improved Euler's method to this initial-value problem.

Table 7.16
Results of improved Euler's method

t_k	y_k
0.	0.
0.25	0.21875
0.5	0.389648
0.75	0.523163
1.	0.627471

Using the analytic solution, we know that the actual value of $y(1)$ is $1 - 1/e$. Therefore, we can compute the error

$$e_4 = |y(1) - y_4| = 0.00464959.$$

(c) If we want an approximation that is accurate to 0.0001, we need an improvement by a factor of

$$\frac{0.00464959}{0.0001} = 46.4959.$$

Since improved Euler's method is a second-order numerical scheme, we expect to get that kind of improvement if we increase the number of steps by a factor of $\sqrt{46.4959}$. In other words, we compute the smallest integer larger than $4\sqrt{46.4959} = 27.2752$. Using $n = 28$ steps, we get the approximate value $y_{28} = 0.63204$ and, consequently, an error $e_{28} \approx 0.00008$.

11. **(a)** The differential equation is separable. Therefore, we integrate

$$\int \frac{1}{y^2}\, dy = \int -1\, dt.$$

We obtain

$$-\frac{1}{y} = -t + c$$

$$y = \frac{1}{t+k}.$$

We determine the value of k using the initial condition $y(0) = 1/2$. Hence, $k = 2$, and the solution to the given initial-value problem is $y(t) = 1/(t+2)$.

(b) Table 7.17 contains the steps involved in applying improved Euler's method to this initial-value problem.

Table 7.17
Results of improved Euler's method

t_k	y_k
0.	0.5
0.5	0.402344
1.	0.336049
1.5	0.288275
2.	0.252281

Using the analytic solution, we know that the actual value of $y(2)$ is $1/4$. Therefore, we can compute the error

$$e_4 = |y(2) - y_4| = 0.002281.$$

(c) If we want an approximation that is accurate to 0.0001, we need an improvement by a factor of

$$\frac{0.002281}{0.0001} = 22.81.$$

Since improved Euler's method is a second-order numerical scheme, we expect to get that kind of improvement if we increase the number of steps by a factor of $\sqrt{22.81}$. In other words, we compute the smallest integer larger than $4\sqrt{22.81} = 19.103$. Using $n = 20$ steps, we get the approximate value $y_{20} = 0.250081$ and, consequently, an error $e_{20} \approx 0.00008$.

13. **(a)** If we want an approximation that is accurate to 0.0001, we need an improvement by a factor of

$$\frac{0.000695}{0.0001} = 6.95.$$

Since improved Euler's method is a second-order numerical scheme, we expect to get that kind of improvement if we increase the number of steps by a factor of $\sqrt{6.95}$. In other words, we compute the smallest integer larger than $20\sqrt{6.95} = 52.7257$.

(b) Using $n = 53$ steps, we get the approximate value $y_{53} = 0.200095$.

(c) Consequently, the error e_{53} is the difference between the actual value $y(2) = 0.2$ and y_{53}. We get $e_{53} = 0.000095$.

15.

17.

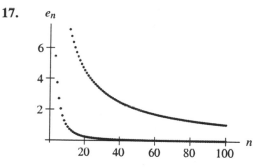

EXERCISES FOR SECTION 7.3

1. Runge-Kutta applied to this initial-value problem yields the points given in Table 7.18. The graph illustrates the results of Runge-Kutta as compared to those of Euler's method, improved Euler's method, and a built-in solver.

Table 7.18
Results of Runge-Kutta

t_k	y_k
0.0	3.000
0.5	8.979
1.0	25.173
1.5	69.030
2.0	187.811

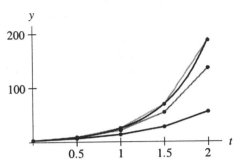

3. Runge-Kutta applied to this initial-value problem yields the points given in Table 7.19. The graph illustrates the results of Runge-Kutta as compared to those of Euler's method, improved Euler's method, and a built-in solver.

Table 7.19
Results of Runge-Kutta

t_k	y_k	t_k	y_k
0.0	0.00000	3.0	2.99645
0.5	1.82290	3.5	2.99882
1.0	2.70058	4.0	2.99961
1.5	2.90368	4.5	2.99987
2.0	2.96803	5.0	2.99996
2.5	2.98935		

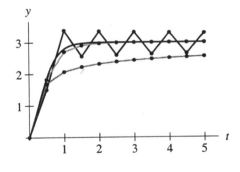

5. Note the relationship between this exercise and Exercise 4.

Runge-Kutta applied to this initial-value problem yields the points given in Table 7.20. The graph illustrates the results of Runge-Kutta as compared to those of Euler's method, improved Euler's method, and a built-in solver.

Table 7.20
Results of Runge-Kutta

t_k	y_k
1.0	2.00000
1.5	3.10456
2.0	3.98546
2.5	4.77554
3.0	5.51352

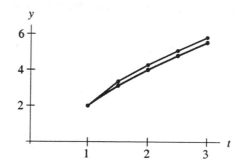

7. **(a)** The results of Runge-Kutta applied to the predator-prey system are given in Table 7.21 and the figure illustrates this computation in the phase plane.

Table 7.21
Results of Runge-Kutta for the
Predator-Prey System

t_k	R_k	F_k
0	1.	1.
1	1.50412	1.91806
2	0.641301	2.48192
3	0.416812	1.62154
4	0.636774	1.08434

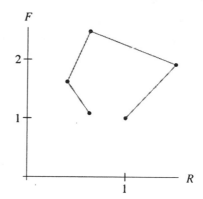

(b)

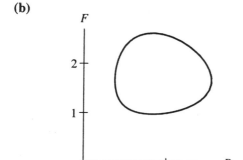

(c)

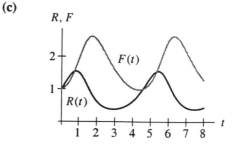

EXERCISES FOR SECTION 7.4

1. Table 7.22 contains the results of approximating the value $y(1)$ of the solution using y_k, where y_k is obtained by applying Runge-Kutta over the interval $0 \leq t \leq 1$ using k steps with single precision arithmetic. Note that we obtain the approximation $y(1) \approx 0.941274$, but we cannot be confident about the next digit.

Table 7.22
Runge-Kutta approximations

k	y_k	k	y_k
2	0.94188476	64	0.94127458
4	0.92484933	128	0.94127458
8	0.94101000	256	0.94127434
16	0.94127089	512	0.94127452
32	0.94127452	1024	0.94127500

3. Table 7.23 contains the results of approximating the value $y(2)$ of the solution using y_k, where y_k is obtained by applying Runge-Kutta over the interval $0 \leq t \leq 2$ using k steps with single precision arithmetic. Note that we obtain the approximation $y(2) \approx 1.25938$, but we cannot be confident about the next digit.

Table 7.23
Runge-Kutta approximations

k	y_k	k	y_k
2	1.33185744	64	1.25938189
4	1.25684679	128	1.25938177
8	1.25911093	256	1.25938165
16	1.25936496	512	1.25938165
32	1.25938094	1024	1.25938201

CHAPTER 8

Discrete Dynamical Systems

EXERCISES FOR SECTION 8.1

1. $x_0 = 0$, $x_1 = -2$, $x_2 = 2$, and $x_3 = 2$. The orbit is eventually fixed.

3. $x_0 = 0$, $x_1 = 1$, $x_2 = e$, and $x_3 = e^e$. The orbit tends to infinity.

5. $x_0 = 0$, $x_1 = 2$, $x_2 = 0$, and $x_3 = 2$. The orbit is periodic of period 2.

7. $x_0 = 0$, $x_1 = 1$, $x_2 = 1$, and $x_3 = 1$. The orbit is eventually fixed.

9. To find the fixed points, we solve $F(x) = x$ or

$$-x + 2 = x.$$

The solution is $x = 1$. For periodic points of period 2, we solve $F^2(x) = x$ or

$$-(-x + 2) + 2 = x.$$

This equation reduces to the identity $x = x$. Therefore, all real numbers (except the fixed point $x = 1$) are periodic points of period 2.

11. To find the fixed points, we solve $F(x) = x$ or

$$x^2 + 1 = x.$$

This equation can be written as $x^2 - x + 1 = 0$ which has imaginary solutions. Alternatively, since

$$F(x) - x = x^2 - x + 1 = (x - \frac{1}{2})^2 + \frac{3}{4} > 0,$$

we see that $F(x) > x$. Thus, the graph of $F(x)$ does not intersect the diagonal and there are no fixed points. For periodic points of period 2, we first compute that

$$F^2(x) = (x^2 + 1)^2 + 1 = x^4 + 2x^2 + 2.$$

Then we solve $F^2(x) = x$ or

$$x^4 + 2x^2 - x + 2 = 0.$$

We already know two solutions of this equation. They are the two imaginary solutions to

$$x^2 - x + 1 = 0.$$

(Any x-value satisfying $F(x) = x$ also satisfies $F^2(x) = x$.) Using long division, we have

$$x^4 + 2x^2 - x + 2 = (x^2 - x + 1)(x^2 + x + 2).$$

Since the roots of the quadratic factor $x^2 + x + 2$ are imaginary also, we have no period 2 points. We also could have used the fact that

$$F^2(x) - x = (x^2 + 1)^2 + 1 - x = x^4 + 2(x - \frac{1}{4})^2 + \frac{15}{8} > 0.$$

Therefore, $F^2(x) > x$ and the graph of $F^2(x)$ does not intersect the diagonal.

13. To find the fixed points, we solve $F(x) = x$ or

$$\sin x = x.$$

Since the graph of $F(x) = \sin x$ only crosses the diagonal at the origin, $x = 0$ is the only fixed point. For periodic points of period 2, consider the function

$$G(x) = F^2(x) - x = \sin(\sin x) - x.$$

Any period 2 point will be a root of $G(x)$. Differentiation yields

$$G'(x) = (\cos(\sin x)) \cos x - 1 < 0 \text{ for } x \neq 0,$$

so that $G(x)$ is strictly decreasing. Since $G(0) = 0$, we have $G(x) > 0$ for $x < 0$ and $G(x) < 0$ for $x > 0$. This means that $x = 0$ is the only root of $G(x)$ and thus there are no periodic points of period 2. Another method would be to use the fact that $|\sin x| < |x|$ for $x \neq 0$. This implies that

$$|\sin(\sin x)| < |\sin x| < |x| \text{ for } x \neq 0.$$

15. For fixed points,

$$F(x) = -2x - x^2 = x,$$

or $x = 0, -3$. For periodic points of period 2,

$$F^2(x) = -x^4 - 4x^3 - 2x^2 + 4x = x,$$

or

$$x(x + 3)(-x^2 - x + 1) = 0.$$

Therefore, the periodic points of period 2 are $x = (-1 \pm \sqrt{5})/2$.

17. From the graph of $F(x) = -e^x$ and $f(x) = x$, there is one intersection and, therefore, there is one fixed point. For periodic points of period 2, consider

$$g(x) = -e^{-e^x} - x.$$

By the graph (or by computing g'), $g(x)$ is zero only once, so this point must be the fixed point and there are no periodic points of period 2.

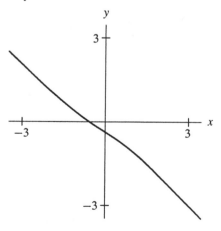

Graph of $y = -e^{-e^x} - x$.

19. The graph of $y = -x$ meets $y = x$ only at $x = 0$, so 0 is a fixed point. Since $F^2(x) = x$, it follows that all other real numbers lie on periodic orbits of period 2.

21. There is a fixed point at $x = 2$; all other points are sent to $x = 2$ by F so they are eventually fixed. Therefore there are no other periodic points.

23. $x_0 = x$, $x_1 = -x + 4$, $x_2 = x$, $x_3 = -x + 4$, and $x_4 = x$. The orbit of any real number is a periodic orbit of period 2, except for $x = 2$, which is fixed.

25. The orbit is $1/6, 1/3, 2/3, 2/3, \ldots$, so $1/6$ is eventually fixed.

27. The orbit is $2/7, 4/7, 6/7, 2/7, \ldots$, so $2/7$ is periodic with period 3.

29. The orbit is $1/8, 1/4, 1/2, 1, 0, 0, 0, \ldots$, so $1/8$ is eventually fixed.

31. The orbit of $6/11$ is $6/11, 10/11, 2/11, 4/11, 8/11, 6/11, \ldots$, so $6/11$ is periodic with period 5.

33. The orbit is $0, 0, 0, \ldots$, so it is a fixed point.

35. The orbit is $1/2, 1, 0, 0, \ldots$, so it is eventually fixed.

37. For fixed points, we must solve

$$F_c(x) = x^2 + c = x$$

or

$$x^2 - x + c = 0.$$

From the quadratic formula we find

$$x = (1 \pm \sqrt{1 - 4c})/2.$$

Thus we need $1 - 4c \geq 0$ or $c \leq \frac{1}{4}$. For $c = 1/4$, we have $x = 1/2$ and this is the only fixed point. For $c < 1/4$, there are two roots and thus there are two fixed points. For $c > 1/4$, there are no fixed points.

39. For periodic points of period 2, we have

$$F_c^2(x) = (x^2 + c)^2 + c = x,$$

or

$$x^4 + 2cx^2 - x + c^2 + c = 0.$$

Since we have fixed points when $x^2 + c = x$, the left-hand side can be factored and one obtains

$$(x^2 - x + c)(x^2 + x + c + 1) = 0.$$

This is zero when $x^2 - x + c = 0$ or when $x^2 + x + c + 1 = 0$. The first factor corresponds to the fixed points, which are present when $c \leq 1/4$. The second factor corresponds to the period two points and has two solutions when $c < -3/4$. At $c = -3/4$ there are two fixed points (roots of the second factor coincide with one of the fixed points).

EXERCISES FOR SECTION 8.2

1. For the fixed points,

$$F(x) = x^2 - 2x = x,$$

or

$$x^2 - 3x = 0.$$

Then, $x = 0, 3$ are fixed points. Differentiation yields

$$F'(x) = 2x - 2.$$

Then, $F'(0) = -2$ and $F'(3) = 4$. Therefore, both $x = 0$ and $x = 3$ are repelling fixed points.

3. For the fixed points,

$$F(x) = \sin x = x,$$

or $x = 0$. Differentiation yields $F'(0) = 1$. For $x > 0$, $x > \sin x$ and for $x < 0$, $x < \sin x$. Therefore, $x = 0$ is attracting by graphical analysis.

5. For the fixed points,

$$F(x) = \arctan x = x.$$

Suppose $f(x) = \arctan x - x$ and differentiation yields

$$f'(x) = \frac{-x^2}{1 + x^2} - 1.$$

Then, $f'(x) < 0$ for $x \neq 0$ and $f'(x) = 0$ only for $x = 0$. $f(x)$ is decreasing for $x \neq 0$. Since $f(0) = 0$, $f(x)$ is tangent to x-axis at $x = 0$. Therefore, $x = 0$ is the only fixed point. Also, $f(x) = F(x) - x > 0$ for $x < 0$ and $f(x) = F(x) - x < 0$ for $x > 0$. Therefore, $x = 0$ is attracting.

7. For the fixed points,

$$F(x) = \frac{\pi}{2} \sin x = x.$$

Then, $x = 0, \pm\pi/2$ are fixed points. Differentiation yields

$$F'(x) = \frac{\pi}{2} \cos x.$$

Then, $F'(0) = \pi/2$ and $F'(\pm\pi/2) = 0$. Therefore, $x = 0$ is repelling, and $x = \pm\pi/2$ are attracting.

9. For the fixed points,

$$F(x) = \frac{1}{x} = x,$$

or

$$x^2 = 1.$$

Then, $x = \pm 1$ are fixed points. Differentiation yields

$$F'(x) = -\frac{1}{x^2}.$$

Then, $F'(\pm 1) = -1$. Since $F^2(x) = x$ for any non-zero x, $x = \pm 1$ are neutral.

11. For the fixed points,

$$F(x) = e^x = x.$$

Suppose $f(x) = e^x - x$ and differentiation yields

$$f'(x) = e^x - 1.$$

Therefore, $f(x)$ is increasing for $x > 0$, decreasing for $x < 0$ and equal to 1 for $x = 0$. $f(x) \neq 0$ for any x and there are no fixed points.

13. $F(0) = 1$, and $F^2(0) = 0$. The period is 2 and since $(F^2)'(0) = 0$, it is attracting.

15. $F(0) = 1$, $F(1) = 2$, $F(2) = 3$, $F(3) = 4$, and $F(4) = 0$. The period is 5 and since $(F^5)'(0) = 2$, the cycle is repelling.

17. $F(0) = 1$ and $F^2(0) = 0$. The period is 2. For $x < 2$ close to zero, $F(x) = 1 - x$. Therefore, any x close to zero is periodic point and the cycle of $x = 0$ is neutral.

19. For fixed points,

$$F(x) = \frac{1}{x} = x,$$

or $x = \pm 1$. Since $F^2(x) = x$ for all other x, these fixed points are neutral. We can also see this by graphical analysis.

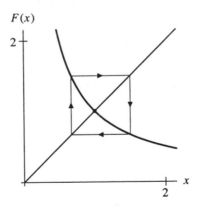

21. For fixed points,

$$F(x) = \tan x = x,$$

or $x = 0$. Since $(\tan x)' = 1/\cos^2 x = 1$ if and only if $x = 0, \pi$, $x = 0$ is the only fixed point with $F'(0) = 1$. By graphical analysis, the fixed point is repelling.

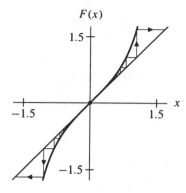

23. For fixed points,

$$F(x) = -x - x^3 = x,$$

or $x = 0$. By graphical analysis, the fixed point is repelling.

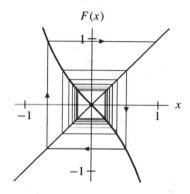

25. For fixed points,

$$F(x) = -e \cdot e^x = x,$$

or $x = -1$. By graphical analysis, the fixed point is attracting.

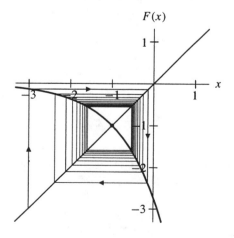

27. For $F(x) = \tan x = x$, there are infinitely many fixed points because $\tan x$ is periodic, and $|\tan x|$ tends to infinity as x tends to $n\pi/2$ for odd integers n. As we saw in Exercise 21, $x = 0$ is repelling. First, note that for $x = n\pi$ for $n \neq 0$,

$$F(n\pi) = \tan n\pi = 0 \neq n\pi.$$

Therefore, $x = n\pi$ for $n \neq 0$ are not fixed points. At the fixed points,

$$F'(x) = \frac{1}{\cos^2 x} > 1$$

because $|\cos x| < 1$. Therefore, all other fixed points are repelling.

29. There is a cycle for T. Try $x = 16/21$. Suppose T has n-cycle, $x_0, x_1, \cdots, x_n = x_0$. Then,

$$|(T^n)'(x_0)| = |T'(x_0) \cdot T'(x_1) \cdots T'(x_{n-1})|$$
$$= 4^n,$$

where n is integer larger than zero. Therefore, the cycle is repelling.

EXERCISES FOR SECTION 8.3

1. For fixed points, we must have
$$F_\alpha(x) = x + x^2 + \alpha = x$$

or $x^2 + \alpha = 0$. Therefore, for $\alpha > 0$, there are no fixed points; for $\alpha = 0$, there is one fixed point; and for $\alpha < 0$, there are two fixed points at $x = \pm\sqrt{-\alpha}$. Differentiation yields

$$F'_\alpha(x) = 1 + 2x$$

and
$$F'_\alpha(\pm\sqrt{-\alpha}) = 1 \pm 2\sqrt{-\alpha}.$$

Therefore, for $-\alpha$ small, $0 < 1 - 2\sqrt{-\alpha} < 1$ and $x = -\sqrt{-\alpha}$ is attracting. Since $1 + 2\sqrt{-\alpha} > 1$, $x = \sqrt{-\alpha}$ is repelling. For $\alpha = 0$, $F_\alpha(x)$ is tangent to $y = x$ from above and, therefore, $x = 0$ is neutral. The bifurcation is a tangent bifurcation.

3. For α slightly smaller than 1, the origin is the only fixed point and it is attracting. For $\alpha = 1$, F_1 is tangent to $y = x$ and 0 is attracting. For $\alpha > 1$, two more fixed points appear and they are attracting for α slightly larger than 1. The origin becomes a repelling fixed point. This is a pitchfork bifurcation.

5. For fixed points,
$$\alpha - x^2 = x,$$

and $x = (-1 \pm \sqrt{1 + 4\alpha})/2$. Therefore, there are no fixed point for $\alpha < -1/4$, there is one fixed point for $\alpha = -1/4$, and there are two fixed points for α slightly larger than $-1/4$. This is a tangent bifurcation. Also, for α slightly larger than $-1/4$,

$$\left| F'_\alpha\left(\frac{-1 + \sqrt{1 + 4\alpha}}{2}\right) \right| = |-1 + \sqrt{1 + 4\alpha}| < 1$$

and

$$\left| F'_\alpha \left(\frac{-1 - \sqrt{1+4\alpha}}{2} \right) \right| = |-1 - \sqrt{1+4\alpha}| > 1.$$

Therefore, $x = (-1 + \sqrt{1+4\alpha})/2$ is attracting and $x = (-1 - \sqrt{1+4\alpha})/2$ is repelling. For $\alpha = -1/4$, $F'_{-1/4}(-1/2) = 1$ and, therefore, $F_{-1/4}$ is tangent to $y = x$ from below at $x = -1/2$. By graphical analysis, $x = -1/2$ is neutral. Nearby orbits on the right of the fixed point tend to the fixed point and nearby orbits on the left tend away from it.

7. This bifurcation is none of the above. When $\alpha = 0$, we have $F_\alpha(x) = 0$ for all x. So all nonzero points are eventually fixed. For $0 < |\alpha| < 1$, the origin is an attracting fixed point and there is a second (repelling) fixed point at $x = (1 - \alpha)/\alpha$.

9. The function T_μ has a single fixed point at the origin when $0 < \mu < 1$. This fixed point is attracting. If $\mu = 1$, all $x \le 1/2$ are fixed points, and each is neutral. If $\mu > 1$, T_μ has two fixed points, at $x = 0$ and $x = \mu/(\mu + 1)$, and both are repelling.

11. The graph of the function $F_c(x)$ is a parabola with minimum at $x = 0$ and $F_c(0) = c$. The function $F_c^2(x)$ is a quartic with local maximum at $x = 0$. Consider the small portion of the graph of F_c^2 defined over the interval $[p_c, -p_c]$, where p_c is the leftmost fixed point of F_c (not F_c^2). This piece of the graph of F_c^2 resembles that of F_c, only upside down. Note that, as c decreases, this piece of graph behaves similarly to that of F_c relative to the diagonal. Thus we expect F_c^2 to undergo a period-doubling bifurcation in this interval, much the same as F_c did.

EXERCISES FOR SECTION 8.4

1.

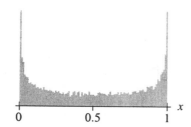
Histogram for seed .3.

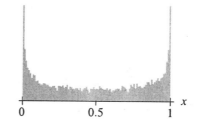
Histogram for seed .3001.

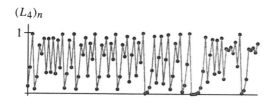
Time series for seed .3.

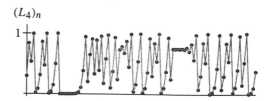
Time series for seed .3001.

3. The image under T of a point of the form $p/2^n$ where p is an integer is of the form $q/2^{n-1}$ where q is an integer. This is because T multiplies by 2 and then subtracts one if necessary to keep the image in the interval $[0, 1)$ (where 1 is not included). Hence, after n iterates, $T^n(p/2^n)$ must be the fixed point 0.

5. The graph of T^n crosses the diagonal in each interval $[i/2^n, (i + 1)/2^n)$ for $i = 0, 1, \ldots, 2^n - 2$. Hence, each of these intervals contains a point of period n (n may not be the least period). We may take n as large as we like, so periodic points of T are dense.

7. The eventually fixed points of T are those whose orbits hit 0. These are precisely the points which have finite binary expansion, that is, points of the form

$$\frac{a_1}{2} + \frac{a_2}{2^2} + \frac{a_3}{2^3} + \ldots + \frac{a_n}{2^n}$$

where $a_1, a_2, \ldots, a_n$ equal zero or one. Using 2^n as a common denominator, we can express such a point in the form $p/2^n$ for an integer p between zero and $2^n - 1$.

9. The points which are fixed point T^n are those whose binary expansion repeats every n summands, that is, those of the form

$$x = \frac{a_1}{2} + \frac{a_2}{2^2} + \ldots + \frac{a_n}{2^n} + \frac{a_1}{2^{n+1}} + \frac{a_2}{2^{n+2}} + \ldots + \frac{a_n}{2^{2n}} + \frac{a_1}{2^{2n+1}} + \ldots.$$

We may choose $a_1, a_2, \ldots$, however we like provided they are not all one (since this corresponds to $x = 1$ which is not in the domain). Hence, T^n has $2^n - 1$ fixed points (and T has $2^n - 1$ points of period n). This can also be seen by looking at the graph of T^n.

11. First form a list of all the possible finite strings of zeros and ones,

$$0, 1, 00, 01, 10, 11, 000, 001, 010, 011, 100, 101, 110, 111, 0000, \ldots$$

then concatenate this list into the binary expansion of a number x_0,

$$x_0 = .0100011011000001010011100101110111 \ldots.$$

Since every finite string appears in this list and application of T shifts the decimal expansion to the left by one place, the orbit of this point will eventually come close to every point in $[0, 1)$.

13. The graph of T^n on the segment $[i/2^n, (i + 1)/2^n]$ is a straight line from zero to one or one to zero depending on whether i is even or odd. In either case, the graph of T^n crosses the diagonal in each such interval. Hence, T^n has a fixed point and T has a periodic point in each such interval. Since n can be taken as large as we like, periodic points are dense.

APPENDICES

EXERCISES FOR APPENDIX A

1. We guess a particular solution of the form $y_p(t) = ae^{-2t}$. Then $dy_p/dt = -2ae^{-2t}$, and we want to find a such that

$$\frac{dy_p}{dt} = y_p(t) + 3e^{-2t}.$$

We obtain

$$-2ae^{-2t} = ae^{-2t} + 3e^{-2t},$$

which is satisfied if $-3a = 3$. Hence, $a = -1$.

The general solution to the associated homogeneous equation is $y_h(t) = ke^t$, so the general solution to the original differential equation is

$$y(t) = ke^t - e^{-2t}.$$

3. We guess a particular solution of the form

$$y_p(t) = a\cos 2t + b\sin 2t.$$

Then

$$\frac{dy_p}{dt} = -2a\sin 2t + 2b\cos 2t,$$

and we want to find a and b such that

$$\frac{dy_p}{dt} = y_p + \cos 2t.$$

We obtain

$$-2a\sin 2t + 2b\cos 2t = a\cos 2t + b\sin 2t + \cos 2t,$$

which is satisfied if

$$\begin{cases} 2b = a + 1 \\ -2a = b. \end{cases}$$

Hence, $a = -1/5$ and $b = 2/5$.

The general solution to the associated homogeneous equation is $y_h(t) = ke^t$, so the general solution to the original differential equation is

$$y(t) = ke^t - \tfrac{1}{5}\cos 2t + \tfrac{2}{5}\sin 2t.$$

5. First we find the general solution of the equation by guessing a particular solution of the form $y_p(t) = ae^{t/3}$. Then $dy_p/dt = (a/3)e^{t/3}$, and we want to find a such that

$$\frac{dy_p}{dt} = -2y_p(t) + e^{t/3}.$$

We obtain

$$\frac{a}{3}e^{-2t} = -2ae^{t/3} + e^{t/3},$$

which is satisfied if $(7/3)a = 1$. Hence, $a = 3/7$.

The general solution to the associated homogeneous equation is $y_h(t) = ke^{-2t}$, so the general solution to the differential equation is

$$y(t) = ke^{-2t} + \tfrac{3}{7}e^{-2t}.$$

To find the solution of the given initial-value problem, we evaluate the general solution at $t = 0$ and obtain

$$y(0) = k + \tfrac{3}{7}.$$

Since the initial condition is $y(0) = 1$, we see that $k = 4/7$. The desired solution is

$$y(t) = \tfrac{4}{7}e^{-2t} + \tfrac{3}{7}e^{t/3}.$$

7. First we find the general solution of the equation by guessing a particular solution of the form

$$y_p(t) = a\cos 2t + b\sin 2t.$$

Then

$$\frac{dy_p}{dt} = -2a\sin 2t + 2b\cos 2t,$$

and we want to find a and b such that

$$\frac{dy_p}{dt} = -y_p + \cos 2t.$$

We obtain

$$-2a\sin 2t + 2b\cos 2t = -a\cos 2t - b\sin 2t + \cos 2t,$$

which is satisfied if

$$\begin{cases} 2b = -a + 1 \\ -2a = -b. \end{cases}$$

Hence, $a = 1/5$ and $b = 2/5$.

The general solution to the associated homogeneous equation is $y_h(t) = ke^{-t}$, so the general solution to the differential equation is

$$y(t) = ke^{-t} + \tfrac{1}{5}\cos 2t + \tfrac{2}{5}\sin 2t.$$

To find the solution of the given initial-value problem, we evaluate the general solution at $t = 0$ and obtain

$$y(0) = k + \tfrac{1}{5}.$$

Since the initial condition is $y(0) = 5$, we see that $k = 24/5$. The desired solution is

$$y(t) = \tfrac{24}{5}e^{-t} + \tfrac{1}{5}\cos 2t + \tfrac{2}{5}\sin 2t.$$

9. **(a)** For the guess $y_p(t) = a \cos 3t$, we have $dy_p/dt = -3a \sin 3t$, and substituting this guess into the differential equation, we get

$$-3a \sin 3t + 2a \cos 3t = \cos 3t.$$

If we evaluate this equation at $t = \pi/6$, we get $-3a = 0$. Therefore, $a = 0$. However, $a = 0$ does not produce a solution to the differential equation. Consequently, there is no value of a for which $y_p(t) = a \cos 3t$ is a solution.

(b) If we guess $y_p(t) = a \cos 3t + b \sin 3t$, then the derivative

$$\frac{dy_p}{dt} = -3a \sin 3t + 3b \cos 3t$$

is also a simple combination of terms involving $\cos 3t$ and $\sin 3t$. Substitution of this guess into the equation leads to two linear algebraic equations in two unknowns, and such systems of equations usually have a unique solution.

11. Let $y(t) = y_h(t) + y_1(t) + y_2(t)$. Then

$$\frac{dy}{dt} + a(t)y = \frac{dy_h}{dt} + \frac{dy_1}{dt} + \frac{dy_2}{dt} + a(t)y_h + a(t)y_1 + a(t)y_2$$

$$= \frac{dy_h}{dt} + a(t)y_h + \frac{dy_1}{dt} + a(t)y_1 + \frac{dy_2}{dt} + a(t)y_2$$

$$= 0 + r_1(t) + r_2(t).$$

This computation shows that $y_h(t) + y_1(t) + y_2(t)$ is a solution of the original differential equation.

13. To find the general solution, we use the technique suggested in Exercise 11. We calculate two particular solutions—one for the right-hand side $t^2 + 2t + 1$ and one for the right-hand side e^{4t}.
    With the right-hand side $t^2 + 2t + 1$, we guess a solution of the form

$$y_{p_1}(t) = at^2 + bt + c.$$

Then

$$\frac{dy_{p_1}}{dt} + 2y_{p_1} = 2at + b + 2(at^2 + bt + c)$$

$$= 2at^2 + (2a + 2b)t + (b + 2c).$$

Then y_{p_1} is a solution if

$$\begin{cases} 2a = 1 \\ 2a + b = 2 \\ b + 2c = 1. \end{cases}$$

We get $a = 1/2$, $b = 1/2$, and $c = 1/4$.
    With the right-hand side e^{4t}, we guess a solution of the form

$$y_{p_2}(t) = ae^{4t}.$$

Then

$$\frac{dy_{p_2}}{dt} + 2y_{p_2} = 4ae^{4t} + 2ae^{4t} = 6ae^{4t},$$

and y_{p_2} is a solution if $a = 1/6$.

The general solution of the associated homogeneous equation is $y_h(t) = ke^{-2t}$, so the general solution of the original equation is

$$ke^{-2t} + \tfrac{1}{2}t^2 + \tfrac{1}{2}t + \tfrac{1}{4} + \tfrac{1}{6}e^{4t}.$$

To find the solution that satisfies the initial condition $y(0) = 0$, we evaluate the general solution at $t = 0$ and obtain

$$k + \tfrac{1}{4} + \tfrac{1}{6} = 0.$$

Hence, $k = -5/12$.

15. To find the general solution, we use the technique suggested in Exercise 11. We calculate three particular solutions—one for the right-hand side $\cos 2t$, one for the right-hand side e^{3t}, and one for the right-hand side e^{-4t}.

With the right-hand side $\cos 2t$, we guess a solution of the form

$$y_{p_1}(t) = a \cos 2t + b \sin 2t.$$

Then

$$\frac{dy_{p_1}}{dt} + 3y_{p_1} = -2a \sin 2t + 2b \cos 2t + 3(a \cos 2t + b \sin 2t)$$

$$= (3a + 2b) \cos 2t + (-2a + 3b) \sin 2t.$$

Then y_{p_1} is a solution if

$$\begin{cases} 3a + 2b = 1 \\ -2a + 3b = 0. \end{cases}$$

We get $a = 3/13$ and $b = 2/13$.

With the right-hand side e^{3t}, we guess a solution of the form

$$y_{p_2}(t) = ae^{3t}.$$

Then

$$\frac{dy_{p_2}}{dt} + 3y_{p_2} = 3ae^{3t} + 3ae^{3t} = 6ae^{3t},$$

and y_{p_2} is a solution if $a = 1/6$.

With the right-hand side e^{-4t}, we guess a solution of the form

$$y_{p_3}(t) = ae^{-4t}.$$

Then

$$\frac{dy_{p_3}}{dt} + 3y_{p_3} = -4ae^{-4t} + 3ae^{-4t} = -ae^{-4t},$$

and y_{p_3} is a solution if $a = -1$.

The general solution of the associated homogeneous equation is $y_h(t) = ke^{-3t}$, so the general solution of the original equation is

$$ke^{-3t} + \tfrac{3}{13}\cos 2t + \tfrac{2}{13}\sin 2t + \tfrac{1}{6}e^{3t} - e^{-4t}.$$

To find the solution that satisfies the initial condition $y(0) = 0$, we evaluate the general solution at $t = 0$ and obtain

$$k + \tfrac{3}{13} + \tfrac{1}{6} - 1 = 0.$$

Hence, $k = 47/78$.

17. Since the general solution of the associated homogeneous equation is $y_h(t) = ke^{-2t}$ and since these $y_h(t) \to 0$ as $t \to \infty$, we only have to determine the long-term behavior of one solution to the nonhomogeneous equation. However, that is easier said than done.

The key is to consider the slopes in the slope field for the equation. We rewrite the equation as

$$\frac{dy}{dt} = -2y + r(t).$$

Using the fact that $r(t) < 2$ for all t, we observe that $dy/dt < 0$ if $y > 1$ and, as y increases beyond $y = 1$, the slopes become more negative. Similarly, using the fact that $r(t) > -1$ for all t, we observe that $dy/dt > 0$ if $y < -1/2$ and, as y decreases below $y = -1/2$, the slopes become more positive. Thus, the graphs of all solutions must approach the strip $-1/2 \le y \le 1$ in the ty-plane as t increases. More precise information about the long-term behavior of solutions is difficult to obtain without specific knowledge of $r(t)$.

19. Since the general solution of the associated homogeneous equation is $y_h(t) = ke^{-t}$ and since these $y_h(t) \to 0$ as $t \to \infty$, we only have to determine the long-term behavior of one solution to the nonhomogeneous equation. However, that is easier said than done.

The key is to consider the slopes in the slope field for the equation. We rewrite the equation as

$$\frac{dy}{dt} = -y + r(t).$$

For any number $T > 3$, let ϵ be a positive number less than $T - 3$, and fix t_0 such that $r(t) < T - \epsilon$ if $t > t_0$. If $t > t_0$ and $y(t) > T$, then

$$\frac{dy}{dt} < -T + T - \epsilon = -\epsilon.$$

Hence, no solution remains greater than T for all time. Since $T > 3$ is arbitrary, no solution remains greater than 3 (by a fixed amount) for all time.

The same idea works to show that no solution can remain less than 3 (by a fixed amount) for all time. Hence, every solution tends to 3 as $t \to \infty$.

21.   (a) Let $y_p(t) = a_1 t + a_2 t^2 + a_3 t^3 + \dots$. Then

$$\frac{dy_p}{dt} = a_1 + 2a_2 t + \dots,$$

and

$$\frac{dy_p}{dt} + y_p = a_1 + (a_1 + 2a_2)t + \dots.$$

Equating this power series with the power series for e^{-3t} (see the statement of the exercise), we get $a_1 = 1$ and $a_1 + 2a_2 = -3$. Therefore, $a_2 = -2$.

(b) If we guess a solution of the form $y_q(t) = ae^{-3t}$, we get

$$\frac{dy_q}{dt} + y_q = -3ae^{-3t} + ae^{-3t} = -2ae^{-3t},$$

so $a = -1/2$ yields a solution. The general solution to the associated homogeneous equation is ke^{-t}, so the general solution to the original differential equation is

$$y(t) = ke^{-t} - \tfrac{1}{2}e^{-3t}.$$

The solution $y_p(t)$ satisfies the initial condition $y_p(0) = 0$. Therefore,

$$y_p(t) = \tfrac{1}{2}e^{-t} - \tfrac{1}{2}e^{-3t}.$$

If we write these two exponentials in terms of their power series, we get

$$y_p(t) = \frac{1}{2}\left(1 - t + \frac{t^2}{2} \mp \ldots\right) - \frac{1}{2}\left(1 - 3t + \frac{9t^2}{2} \mp \ldots\right).$$

Note that, when these power series are combined, we get $a_0 = 0$, $a_1 = 1$, and $a_2 = -2$.

23.   **(a)** Let $y_p(t) = a_1 t + a_2 t^2 + a_3 t^3 + \ldots$. Then

$$\frac{dy_p}{dt} = a_1 + 2a_2 t + \ldots,$$

and

$$\frac{dy_p}{dt} + 2y_p = a_1 + (2a_1 + 2a_2)t + (2a_2 + 3a_3)t^2 + \ldots.$$

Since

$$e^{-2t} = 1 - 2t + 2t^2 - \frac{8t^3}{3!} \pm \ldots,$$

we must have $a_1 = 1$, $a_2 = -2$, $a_3 = 2$ (not requested in the statement of the exercise), ...

(b) If we guess a solution of the form $y_p(t) = ate^{-2t}$, we have

$$\frac{dy_p}{dt} + 2y_p = a(1 - 2t)e^{-2t} + 2ate^{-2t}$$

$$= ae^{-2t},$$

and $y_p(t) = ate^{-2t}$ is a solution if $a = 1$.

We need to compute the first few terms of the Taylor series for te^{-2t}. Using the series for e^{-2t} given in part (a), we get

$$te^{-2t} = t\left(1 - 2t + 2t^2 - \frac{8t^3}{3!} + \ldots\right)$$

$$= t - 2t^2 + 2t^3 - \frac{8t^4}{3!} \pm \ldots.$$

Note that the coefficients for this series are $a_0 = 0$, $a_1 = 1$, $a_2 = -2$, $a_3 = 2$, ...